SOTHIS PRESS

Sothis Press aims to completely reimagine the public domain book with an emphasis on universal access to captivatingly designed print editions. We believe that technology will enable every single book ever published to be perpetually accessible in beautiful print form. Sothis is beauty in ubiquity.

www.sothis.press
1 Waterhouse Square, London, ECIN 2ST

Epping Forest

With Chapters on Forest Management,
Geology of the District, Prehistoric Man
and the Ancient Fauna, Entomology,
Pond Life, and Fungi of the Forest

Edward North Buxton

SOTHIS PRESS

ISBN 978-1-0368-0511-1

This edition is published by
Sothis Press © 2023. The company
is registered in England and Wales.
The Company Number is 14900389.

EPPING FOREST

BY

EDWARD NORTH BUXTON

VERDERER

SIXTH EDITION REVISED

WITH CHAPTERS ON FOREST MANAGEMENT, GEOLOGY OF THE DISTRICT, PREHISTORIC MAN AND THE ANCIENT FAUNA, ENTOMOLOGY, POND LIFE, AND FUNGI OF THE FOREST

LONDON: EDWARD STANFORD
12, 13, & 14, LONG ACRE, W.C.
GEOGRAPHER TO THE KING
1901

PREFACE TO FOURTH EDITION

As the public have absorbed several thousands of this little guide, I am tempted to reissue it with some emendations and additions. Those whose duty it is to manage the Forest are the target of numerous criticisms and suggestions. I have therefore written a chapter on the Management of the Forest, in which I have examined the debatable questions of policy and principle upon which the well-being of the woodland and its continued natural character depend.

The number of those who study some branch of natural history within the precincts of the Forest is largely on the increase. I have endeavoured to give some directions for the guidance of these students, and I have been so fortunate as to secure the co-operation of Mr. T. V. Holmes, Mr. Henry Woodward, Mr. W. Cole, and Mr. M. C. Cooke, in those branches of science of which they are distinguished exponents.

Since the first issue of this book in 1885, Mr. Fisher's learned and exhaustive work on the Forest of Essex has appeared. To its pages I am indebted, and to them I would refer all who desire to study those ancient forestal customs upon which I have but briefly touched. I am also indebted to my neighbour Mr. W. C. Waller, who has laboriously unearthed and translated many ancient documents which throw light on those laws and practices which have been instrumental in handing down to us our superb fragment of natural forest.

E. N. BUXTON.

PREFACE

THE idea of writing a Guide to the Forest occurred to me when I observed how small a percentage of our summer visitors ever venture far from the point at which they are set down by train or vehicle. This is hardly a desirable state of things; and as steam merry-go-rounds and "five shies for a penny" can be enjoyed with equal facility in London, it seems a pity not to encourage, as far as lies in one's power, a more enterprising spirit. It appears that this reluctance to enter the thicket springs, not from indifference to the attractions of the Forest, but solely from a dread, not unnatural to those unaccustomed to the country, of losing the way; and I hope that clear directions how to find, and follow, the most beautiful and interesting routes will be appreciated. I trust I am not egotistical in thinking that, as I have lived all my life in one or other of the Forest parishes, and for many years have been in the constant habit of exploring its inner recesses, the public cannot have a better adviser than myself in this matter.

I am aware that some excellent cheap Guide-books to the Forest have been published; but they do not enter with sufficient minuteness into topographical details to serve as a handbook to the stranger who desires to penetrate the wilder parts of the wood. An essential part of my scheme is that the descriptions should be accompanied by correct maps of a scale sufficiently large to facilitate identification of the leading features. Taking the official map published by the Conservators, which is of the scale of 3 inches to the mile, I have divided it into sections of a convenient size, which, after careful correction, have been engraved with great care by Mr. Stanford, to whose manager, Mr. Bolton, my acknowledgments are especially due.

To the Itinerary I added an historical chapter, believing that those who use the Forest would wish to know the influences which, centuries ago, made it what it is, and how it came to be rescued for their enjoyment. For some very useful suggestions bearing on this part of the work I have to thank Sir Arthur Hobhouse and Mr. Robert Hunter, than whom no men are better acquainted with the legal aspects of the question.

I find that the most intelligent interest is taken in our woodlands by those who there pursue their studies in one or other of the branches of Natural History. Many of them are of humble station; but as members of Societies, which are very numerous in London, of botanists, ornithologists, fungologists, and microscopists, they are enthusiastic searchers into nature. I have therefore added some chapters on the different forms of life which they may expect to find in the course of their rambles. In this task I have been much assisted by such able naturalists and experienced observers as Professor Boulger, Mr. J. E. Harting, Mr. English, and others.

Mr. Heywood Sumner has embellished the text with some charming drawings, some of which I hope will incidentally be of value in helping to identify the features and landmarks by which I have sought to point out the way.

To Mr. J. Wycliffe Taylor and to Mr. Moon I am also indebted for some beautiful sketches, most truthful to nature, in the one case of birds, and in the other of flowers.

I have also to thank my brother-verderers and several of my neighbours for many valuable suggestions.

E. N. BUXTON.

CONTENTS

PAGE

HISTORY 1

TOPOGRAPHY 22

Route A—Leytonstone to Wanstead Park and Forest Gate 23

Route B—Leytonstone to Woodford or Buckhurst Hill 25

Route C—Wood Street Station to Chingford . 26

Route D—Ponder's End Station, on the Great Eastern Main Line, to Chingford or Loughton 29

From Chingford to High Beach 31

Route E 32

Route F 33

Route G 33

Route H 35

Route I 35

Route J—Loughton to High Beach 35

Route K—High Beach to Loughton 36

Route L—Buckhurst Hill or Loughton to Epping 37

Route M—Loughton to Theydon by Baldwin's Hill and Oak Hill 38

Route N—High Beach to Epping by the Verderer's Path and Ambresbury Banks 39

Route O—Theydon to Waltham by Debden Green 40

Route P—Loughton to High Beach by Baldwin's Hill 41

Route Q—Theydon Station to Waltham by Copped Hall Green 42

Route R—Theydon to Hangboy Slade . . . 43

Route S—Theydon to Epping 44

Route T—The Lower Forest 44

PAGE

OBJECTS OF INTEREST WITHIN AND AROUND THE FOREST—
- Wanstead Park 45
- Ambresbury Banks and Loughton Camps . . 47
- Greensted Church 49
- Old Chingford Church 51
- Queen Elizabeth's Lodge 51
- The Abbey of the Holy Cross of Waltham . . 51

ANIMALS 56
REPTILES 66
BIRDS 67
ENTOMOLOGY 83
POND LIFE 98
PREHISTORIC MAN 104
TREES 109
FLOWERING PLANTS 118
MOSSES 130
FUNGI 133
GEOLOGY 143
MANAGEMENT 150

APPENDIX—
- Monk Wood, in Loughton 171

ILLUSTRATIONS

	PAGE
Border from the Arundel Psalter, A.D. 1329	1
Monk Wood	15
Fairmead, from Warren Hill	31
Cuckoo Pits	34
Waltham Abbey	52
Roe-Deer, Red-Deer, Fallow-Deer	57
Stoat	64
Sparrow Hawk	68
Jay	70
Green Woodpecker	71
Kingfisher	72
Hawfinch	73
Great Tit	75
Longtailed Tits	76
Woodpigeons	77
Lapwings	78
Herons	79
Waterhen	80
Wild Duck	81
High Beach	110
Honeylane Hill	112
Sundew	121
Dewberry	123
Teasel	124
Geological Section	145

MAPS AND PLANS

INDEX MAP	*To face page*	22
MAP A	,,	24
MAP B	..	26
MAP C	..	28
MAP D	..	36
MAP E		44

BORDER FROM THE ARUNDEL PSALTER, A.D. 1329.

EPPING FOREST

ITS HISTORY

"This is the forest primeval."—*Evangeline.*
"I 'a stubbed Thornaby Waäste."—*Northern Farmer.*

A VERY brief account of the history of the Forest, showing how the primeval woodland was set aside for the pleasure of the King, and so was preserved with little change from century to century, until it came to be dedicated as an open space for the enjoyment of King Demos, will not be out of place here. The forest, as we know it, is probably but little altered in character from what it was in the time of the Druids, but its boundaries are sadly curtailed, and that which is now devoted to the recreation of the people and the pasture of the commoners' cattle is but a fraction of the ancient Royal Forest.

As regards its ancient limits, it appears that in Saxon times, and for some centuries later, a large proportion of the county of Essex,—according to Gibson "as far as the sea,"—was "waste," *i.e.* natural woods, open heathery spaces, and grassy commons. When the whole of London was no bigger than Colchester is now, and the population was everywhere sparse, there was not the demand for cultivated ground which arose at a later period. Even as late as the time of Camden, who wrote at the end of the 16th century, we learn that "Near the Ley . . . spreads out a chase of vast extent, full of game, the largest and fattest deer in the kingdom; called heretofore, by way of eminence, the Forest of Essex, now Waltham

Forest, from the town of Waltham, in Saxon Wealdham, *i.e.* a dwelling in the woods." But that it was not all barren we learn from Norden, who, writing at about the same period, said, "This shire is most fatt, fruitfull, and full of profitable thinges. . . . This shire seemeth to me to deserve the title of the Englishe Goshen, the fattest of the lande: comparable to Palestina that flowed with milke and hunnye. But I cannot comende the healthfulnes of it: and especiallie nere the sea coastes . . . which gave me a moste cruell quarterne feuer."

In the reign of King John the grievances of the people against the Forest laws became intolerable, and he was compelled to limit the Royal Forest to what was known as the Forest of Waltham, and although he sought to evade the consequences of this concession, it was subsequently confirmed by Edward I. Its boundaries extended from the river Lea on the west, to the Romford Road on the east; and its northern limit passed through Nazing, Abridge, and Havering-atte-Bower. Three centuries later King Charles I. attempted again to enlarge, not only this Forest, but pretty nearly all the Royal Forests. His object was not so much to extend his hunting grounds as to extort exorbitant fines from those who held lands within the enlarged boundaries. He also sought to raise money by a scheme for the total disafforestation and sale of Waltham and other Forests.[1] These attempts caused great discontent, and were met

[1] That this scheme was at one time nearly realised is shown by the existence among the State Papers preserved in the Public Record Office of a list of claims of owners, upon a contemplated disafforestation of Waltham Forest. Along with this is another State Paper, which shows that in the case of Gaultres Forest the disafforestation was actually accomplished "in pursuance of His Majesty's Royal resolution, as well for raising moneys for supply of his weighty affairs, as for the increase of his revenue"; and we learn further that His Majesty received £20,000 for his share, and also that "the said inhabitants by strong hand do make havoc and insufferable spoil of the woods, drive and keep their cattle upon His Majesty's said part of land, and being disturbed by the purchaser's agents, do make pound-breaches, rescue their cattle, and beat and wound such as disturb them."

with determined opposition, especially in the county of Essex, when my sturdy fellow-countymen were guilty of some "fashious actions" and even had the impudence to hold a "conventicle" "in the very brake where the king's stag should have been lodged for his hunting the next morning." The result was that in the first session of the Long Parliament an Act was passed to fix the Forest boundaries, and a perambulation was made which showed that Waltham Forest comprised 60,000 acres. It must not be supposed, however, that the whole of this wide area was woodland and heath. In the time of the Tudors there is reason to believe that four-fifths were under some sort of cultivation, but at the same time the whole was strictly subject to the Forest laws and regulations for the time being; and the interests of the cultivator, of the commoner who turned his cattle out to graze on the Forest, of the owner of the soil, and of the public, were subordinate to the sporting rights of the Crown.

Previous to the time of Henry VIII. nearly the whole of the ownership of the soil, subject to the sporting rights of the Crown, was vested in the powerful religious bodies of that day; such as the Abbey of Waltham, of which some account is given at page 51, which had received them as grants from various monarchs, pious or rapacious, in return for benefits of a spiritual or substantial character; but other rights and favours were also constantly granted to ecclesiastics, among whom the Abbess of Barking seems to have been especially favoured, as the following extract shows: "We command you to allow the Abbess of Barking her reasonable estovers in her wood at Hainault for her firing, her cooking, and her brewing, if she has been accustomed so to do in the time of our Lord King John our father; also to permit the same abbess to have her dogs to chase hares and foxes within the bailiwick if she was accustomed to have them in the time of our aforesaid father." At the time of the Reformation these rights reverted, or were reannexed, by the

Crown. Later on they were again granted to various persons for services received, with the exception of those which related to the section which lay to the north-east of the river Roding, called the Forest of Hainault, which had belonged to the Abbey of Barking, and which unfortunately, as I shall presently show, remained in the hands of the Crown.

Our business, however, lies chiefly with that part known as Epping Forest, which, having been divided by the Crown among many favoured persons, descended from them, by patrimony or purchase, to the eighteen lords of manors who held them until they were recently acquired by the Corporation of London.

For the proper understanding of the history of the Forest it is desirable to divide it into three periods, during which three distinct influences or ideas prevailed as to the use to which it should be put. During the first, which lasted at least from pre-Norman times until the eighteenth century, the maintenance of the sporting rights of the sovereign was paramount, and to this every other interest was subordinate. About the time that this function of the Forest had lapsed by disuse, though the right still existed, a new idea began to lay hold of the public mind, and a very true one within proper limits, that "he is a benefactor who makes two blades of grass grow where one grew before," and that to bring waste land into cultivation was a public duty. Subsequently the growth of population forced upon the minds of far-sighted people the new truth that fresh air and recreation are not less necessary than food, and of peculiar importance to those who live in great cities.

Each of these influences had a notable effect on the Forest, and combined to make it what it is. I propose to trace briefly the condition of things during the three periods when these separate currents of opinion prevailed.

The Norman kings, as well as their predecessors and successors, were great hunters, and of one of them—William the Conqueror—the Saxon Chronicle says that "he loved the Great Game as if he had been their Father." Being besides men of arbitrary will, and very great political power, they increased the stringency of the previously existing Forest laws. They assigned as Royal Forests many wooded districts in various parts of the country, and enlarged the boundaries of others. It must not, however, be supposed that they were then for the first time turned into "waste" land. According to Lord Coke the great majority of them had from immemorial antiquity been forests, as we understand the word, but not necessarily subject to Forest laws. The Royal Forests have been defined as "certain territories of woody grounds and fruitful pastures, privileged for wild beasts and fowls of forest chase and warren, to rest and abide there in safe protection of the king for his delight and pleasure." They were not enclosed, but were "meered and bounded with unremoveable marks, meers and boundaries," that is, natural landmarks such as hills, rivers, and trees, and boundary stones, some of which have been sought out and discovered in recent years. One old writer, enlarging on the use of the Forests, says they "are for the profit and strength of the kingdom, for the forests are the ships' nurseries of timber . . . likewise of the king's ships, which chase from robbing of it the wasps and hornets, I mean the pirates and the greater enemies"; but he lays most stress on the fact that they are the "chief object of the king's princely delight." And he especially commends Waltham and Windsor Forests on account of their convenience for the "entertainment of foreign princes and of their agents and ambassadors, because the nearness of these two forests unto the city doth much add to the pleasure of them."

Although in some cases, such as the New Forest, cultivated lands were laid waste to gratify the royal pleasure, villages and enclosures were often allowed to exist within the limits; but it is important to observe that they and their inhabitants were *subject to the full stringency of the*

Forest laws. One end which the kings had in view, in enacting these laws, and which embodied a policy natural to a conquering people, and found in all the countries subjugated by the Germans, was to keep weapons out of the hands of the people. Another reason was the extortion which severe laws enabled the Crown to practise. But the *primary object*, to which everything else had to bow, was the preservation of the Forest animals, especially the deer—the King's right of "vert and venison," as it was called. Not only were these animals forbidden to be killed, under penalties of mutilation and even death, but the fences were kept down to such a height that a doe with her fawn could readily jump them,[1] and the owners could not even drive the deer from their crops, on which they fattened; nor could new houses be erected, because of the "increase of men and dogs and other things, which fright the deer from their food"; or the cultivation changed, or trees cut down in enclosed lands without special leave granted by the Forest Courts. Dogs were "*expeditated*," that is, three claws of the fore-feet were cut off close to the ball of the foot to prevent their chasing the deer, and one writer says that only such dogs were allowed "as would go through the Lord Vesci's stirrup, who was Justice in Eyre in Edward II.'s time." The severity of these penalties and restrictions, originally enacted by King Canute, was somewhat mitigated by a charter, "Carta de Foresta," which formed part of Magna Charta extorted from King John, who had been forced to abandon large tracts of Royal Forest; but in their main lines they continued to be recognised and enforced for many centuries since his time. The following are specimens of extracts from the clauses of these laws, lenient by contrast with some others which might be quoted, and fairly show their character:—

"For the offence of Venison or hunting, not unworthely haue bene accounted amongest the

[1] This condition has had an important legal bearing on the fate of the Forest, even in modern times.

greatest offences of the Forest euen of auncient time. But the offence of Vert (except it be for the breach of our Roiall free chace) it is so little, and of so smal an account, that this our Constitution or Law doth scantly respect the same. . . ."

"If any freeman shall chase away a Dere, or a wilde beast out of the Forest: whether the same be done by chaunce, or of a set purpose, so that thereby the wilde beast is forced by swift running to lyll out the tong, or to breathe with his tong out of his mouth: he shall paie to the king ten shillings amends for the same offence: but if he be a seruile person, then he shall double the same recompence: but if he be a bondman, then he shall lose his skinne. . . ."

"If a greedy rauening dogg shall bite a wild beast, then the owner of the same dogg shall yeeld a recompence to the king for the same according to the valew of a freeman which is twelue times a hundred shillings. If a Roiall beast shall be bitten, then the owner of the dogg shal be gilty of the greatest offence. . . ."

These laws were put in force by Forest Courts maintained for the purpose of resisting encroachments, and protecting the deer and the vert. These were the *Court of Attachment*,[1] or forty-day court, presided over by four verderers, who were chosen by the freeholders of the county, hearing cases against offenders, in the first instance, and dealing at once with offences when the damage was not more than fourpence; the *Court of Swainmote*, also presided over by the verderers, but assisted by a jury of freeholders in the Forest,[2]

[1] Some speak also of a *Cheyne* Court for the expeditation of dogs.

[2] In the Stowe Collection, recently acquired by the British Museum, there is the transcript of a charge delivered to the jury and officers constituting the Swainmote Court in 1634, and the Court seems to have been presided over on that occasion by one of the judges. It is probable that he was a creature of Charles I., and that this unusual step was taken to enable him to allay the rising irritation, felt at that time against the Forest Laws, and the strained interpretation which Charles sought to place upon them. A large part of the charge is occupied by an elaborate vindication of the king's rights.

who tried offenders, but if found guilty, sentence was reserved for the highest court, or *Justice seat*, held at much longer intervals, and presided over by the Chief Justice in Eyre. In addition all general questions of right were tried by a jury of freeholders before the same authority, which from time to time issued orders for the regulation of the Forest. The following quotations of orders from the Court Rolls as recently as the beginning of the eighteenth century are fair illustrations of the workings of these courts:—

The licenses to shoot small game in the Forest usually included an authorisation "to take all guns, dogs, and other engines and instruments, wherever the sportsman should find them, from all mean and disorderly persons who are not permitted by law to have and use such things" (Dec. 18, 1713).

On Jan. 7, 1717, one of the under-keepers "presents 3 brase of deer rascally found dead."

On June 17, in the same year, the court "ordered that ye toyles brought to this court by Samuel Heybourne be burnt or cutt in pieces dureing the sitting thereof."

On July 25, 1719, the court ordered "that ye Beadle of the Forest give notice to ye poor people adjoining to ye Forest to keep their geese near to their houses and not let them ramble upon ye Forest otherwise shoot them."

On Nov. 22, 1733, it was ordered "that the severall keepers of the Forest DO hinder the poor people from gathering up the Dung and laying it in heaps in ord^r.^ to carry it away upon their owne lands."

On Oct. 14, 1749, it was ordered "that the keepers do take up and impound all Asses going upon the Forest."

On Aug. 28, 1750, it was ordered "that the gun taken by Joseph Mason from Thos. Gill be given to Mr. Gore a farmer at Waltham to guard him in his journeys to and from London."

The large room in Queen Elizabeth's Lodge was designed and used for the holding of the lesser courts, which made such orders as the above. Licenses or "Deputations," as they were called, to shoot, red-deer and fallow excepted, were granted to individuals of position, subject to the proviso "that he do use the liberty with that moderation which is fitting." These were continued to quite recent times, and two or three holders of licenses lived until quite recently. The well-known figure of Mr. Henry Vigne was one of them. He kept harriers on the Forest for over sixty years; and to the end of his life used to shoot in it on bye-days.

For the watching of the Forest, and for the enforcement of the decrees, there were, besides the above-named *verderers*, thirteen master keepers, who were gentlemen of landed estate, and a number of *foresters*, *agistors*, and *regarders*, who attended upon His Majesty's vert and venison, reported encroachments, and detained and charged offenders; *woodwards*, who, as the name implies, had a care for the timber; and *reeves*, who marked the cattle of those who had commoning rights, and impounded stray beasts. The functions of the reeves are not less important at the present day than they were at the time of the Saxons. Over all these officials was an hereditary *Lord Warden*, whose duty it was to maintain the forest unimpaired for the king's pleasure, and who enjoyed the following perquisites:—"Of every covert and hedgerow to be sold, of every shilling one penny; and of every wood to be sold, the second best oak; and of the buyer and seller of every such wood, one bow to one broad arrow; and one penny of every shilling of the seller and buyer of every such wood upon the sale of it."

The nature of the functions originally entrusted to the chief officers of the Forest is well shown by the following quaint rhyming charter of appointment by Edward the Confessor; but this refers to the earliest period of which we have any documentary evidence relating to the Forest laws:—

Ich, Edward Koning,
Have yeven of my forest the keping

Of the hundred of Chelmer and Dancing
to Randolph Peperking and his kindling
Wyth heorte and hynde, doe and bocke,
Hare and foxe, Catte and brocke,
Wylde fowel with his flocke
Partrich, fesant hen, and fesant cocke,
With green and wylde stob and stocke
to kepen and to yeinen-by al her might,
Both by day and eke by night.
And hounds for to hold
Good and swift and bold
Four greyhounds and six racches
For hare and foxe and wilde cattes;
And therefore Iche maac him my booke
Witness the Bishop Wolston,
And book ylered many on
And Swein of Essex, our Brother
And teken him many other;
And our steward Howœlin
That by sought me for him.

It may be asked how it was that any were found to live within the bounds of the Forest, where they were subject to the pressure of these Forest laws; but the answer lies in the fact that they received very valuable compensation in the commoners' rights, which they possessed from ancient times, of depasturing their cattle on the waste; of the *pannage* of pigs; and of lopping the trees for fuel; and, in addition to these recognised advantages, there were, if report speaks truly, some illicit ones, in the shape of an occasional chance of a fat buck to the man who was not afraid to run the risk. Certain it is that Woodredon Farm, which was then, and is now, a favourite resort of the deer, commanded an excessive rent on account of the facilities which it offered for poaching.

The right of commoning was limited to horses and "neat beasts" [cows] "couchant and levant," and extended to the whole year except during the fence month, commencing the 25th June, when the cattle were required to be withdrawn, on account of the does and newly-born fawns; but since the Crown allowed their rights to fall into abeyance the close time has not been enforced.

In most royal forests there is another close time, called *Winter Heyning*. This extends through the winter months, and is for the purpose of reserving the food, which is then scarce, for the deer; but I cannot find any record of this having been enforced in Epping Forest.

The same persons who enjoyed the privilege of commoning cattle had also the right of *pannage*, *i.e.* of turning out their pigs at Michaelmas to eat acorns and beech-mast. The amount of *pannage* varied in different seasons, but it was always a valuable right, woods being often considered rather by the number of pigs they could feed than by the timber.

The right of lopping was enjoyed in some manors by persons to whom special assignments had been made of so many acres, and in Loughton the right extended to all the householders of the parish. This right had an important influence on the fate of the Forest, as I shall presently show. The object of the forestal laws being to maintain the *status quo*, it is not surprising that the condition of the Forest, as I have described it, remained without material alteration for many centuries; but I now come to what I have called the second period—the utilitarian age—when it was sought to absorb the waste for cultivation and building. The earliest expression of the idea that this was beneficial which I can find is contained in Harrison's "*Description of England*," written in the 16th century—"Certes, if it be not one curſe of the Lord, to haue our countrie conuerted in ſuch ſort from the furniture of mankind, into the walks and ſhrowds of wild beaſts, I know not what is anie. How many families alſo theſe great and ſmall games haue eaten vp, and are likelie hereafter to deuoure, ſome men may coniecture * * * they ſhall ſaie at the laſt, that the twentith part of the realm is imploied vpon deere and conies alreadie, which ſeemeth verie much, if it be dulie conſidered of." The development of this idea is the dark page in my history, when the Forest was despoiled of many a beautiful glade. Let us pass over the bad quarter of an hour as rapidly as possible.

The machinery of the Forest laws existed in full force until near the end of the eighteenth century, and was effective in preventing encroachments. Such enclosures as were allowed were all recorded in the Court Rolls, and were exceedingly minute both in number and in quantity. But Mr. Wellesley Pole, who, in his wife's right, became Lord Warden, saw that more profit was to be made in breaking his trust than in keeping it, and by refusing to support the authority of the verderers, but on the contrary by his example encouraging persons to defy the Forest law, and in other ways, designedly brought the Forest laws into abeyance so that he might himself infringe them with impunity. This he proceeded to do by selling the rights he was appointed to guard. At the final settlement of the Forest question, the office was abolished by Act of Parliament, which directed that compensation should be given to its then holders, the trustees of Lord Mornington, for being deprived of this hereditary trust. They claimed £1000, and were awarded £300 in the arbitration.

The use for sporting purposes by the Crown having all but ceased, and its chief representative seeking to turn his office to profitable account, it is not surprising that the Forest laws were laxly administered, and that the machinery became rusty. Concurrently the growth of London increased the demand for produce of all kinds, and, while the supply was artificially limited by protective laws, put a premium upon the extension of cultivated ground, and enhanced the value of land for building. The term "waste" of the Forest seemed to imply that it was of no value to anybody while it remained in that condition, and tempted those who had the power, or thought they had, to turn it to account. Naturally those parts adjacent to the metropolis were the first to fall into the hands of speculators, and a glance at the map shows that the southern half of the Forest, even now that such large restitutions have been made, presents a sadly tattered appearance. It was, however, not till comparatively recent times that the Forest Courts began to be generally disregarded and defied. The tradition of their terrors still remained; and, though small encroachments were made from time to time, it was not until the middle of the nineteenth century that wholesale enclosures began to be effected.

The office of Woods and Forests was largely responsible for the mutilation which ensued. In the case of Hainault Forest, which lay to the east of the River Roding, the Crown were owners, not only of the sporting rights, but of the manorial rights and the soil as well; and, when the Crown rights were commuted, it was vested in the above-named officials for the benefit of the people at large. They have unfortunately taken an exceedingly narrow view of their trust, and have acted on the short-sighted doctrine that they were to show the best balance-sheet which they could at the moment. Their policy was thus expressed by the Chief Commissioner, the Hon. K. Howard, who, speaking at the moment of the New Forest, said, "We want to get as much as we can out of the Forest"; and what they expected to get is shown in the following recommendation by one of his assistants in the New Forest, of a large enclosure by the Crown, "because, exclusive of other advantages, by so doing, *all the best pasture would be taken from the commoners, and the value of their rights of pasture would thus be materially diminished, which would be of importance to the Crown in the event of any such right being commuted.*" Neglecting the future value of open spaces near great cities, they recommended —and the House of Commons unhappily endorsed the proposal in 1851—the complete destruction of the woodlands of Hainault, and its conversion into arable land. A contract was made with a manufacturer of steam ploughs, who, attaching his anchors to the soil, dragged up by the roots the old oaks, including the Fairlop Oak of ancient memory. The whole operation was concluded in six weeks, and public opinion, even if it had been then ripe for a protest, had no time to act before the work of desecration was complete. Thus the Hainault division of the Forest disappeared for ever, with the exception of a small and very pretty wood by Lambourne, belonging to Colonel Lockwood.

In the case of Epping Forest, governed, as at Hainault, by the same narrow policy, unaccompanied by foresight or common sense, but possessing here only the Crown's forestal rights, and not the manorial rights or the soil, the Commissioners proceeded to offer these rights for sale to the various lords of manors for about £5 an acre; nay, almost to compel the purchase by threats of selling them to others, and by the inducement that their purchase by the lords would give the absolute right to enclose. We cannot greatly blame the latter if, after acquiring the Crown rights under these circumstances, they took an exaggerated view of their powers of enclosure. They now had only the commoners to deal with, and they believed that if each could come to an agreement with this class in his own manor, he would have satisfied every claim. On the face of it, this did not seem an unreasonable view, and though we now know that it was erroneous in law, all we need say, or can say, is that they were badly advised. At the same time the extension of railways, by enabling those who had business in the city to reside beyond its limits, enormously increased the demand for, and the value of, land in the suburbs, and offered a new temptation. Enclosures of large blocks, often of several hundred acres in extent, followed one another with alarming rapidity. In 1850 the Forest comprised almost exactly 6000 acres. During the ensuing twenty years just half this area was surrounded by fences and partly built upon.

While this was proceeding a new spirit arose, or was stimulated by the impending danger, and I am happy to think that I am now entering on the third period, when a truer and a juster view of the needs and rights of the public began to prevail. Opinion was at the same time influenced by the fact that, as the markets of the whole world had been thrown open to us by Sir Robert Peel, the importance of laying every available acre under the plough was diminished.

We now reach the great legal contest which lasted fifteen years, and resulted in the preservation of the 5500 acres which the public now enjoys.

A society, called "The Commons Preservation Society," was formed for the purpose of resisting encroachments on this and other open spaces.

MONK WOOD.

Mr. Shaw-Lefevre, to whose sustained zeal throughout the prolonged contest which I am about to summarize the final success was largely due, was the first chairman, and it has included such names as Mr. Cowper-Temple, afterwards Lord Mount-Temple, Mr. W. H. Smith, Mr. John Stuart Mill, Sir Charles Dilke, Mr. Fawcett, Mr. Charles Buxton, and Miss Octavia Hill—all of whom took an active part in its deliberations. Such a body had, as was to be expected, great parliamentary influence. Among other steps that were taken at its instigation, a Committee of the House of Commons was appointed, who reported in 1863 that the Forest was being destroyed, and recommended that the forestal rights of the Crown should be enforced where they had not been sold.

Two years later another committee was appointed to enquire into all the open spaces round London. The view of the lords of the manor, which was urged upon these committees, was that nothing stood between each lord and enclosure on his own manor but the rights of the commoners of that manor, which rights in several of the manors had either been compensated or surrendered. On the other hand, the opponents of this view maintained that the common rights still existed, and even if little used, were sufficient to resist enclosure. This committee repeated the recommendation with regard to the Forest which had been made by the previous one; but, though an Act was passed providing for the regulation of commons, no steps were taken by the Government of the day with regard to the Forest, and in default of this the party of enclosure gained courage to assert what they believed to be their rights by further large enclosures. Within a comparatively brief period nearly 3000 acres were thus surrounded by fences. Public opinion began to be aroused, but the first overt act of resistance was committed by a labouring man—one Willingale—who persisted in asserting his ancient right of lopping in Loughton Manor, as his forefathers had done. It is held locally that this right was granted by Queen Elizabeth to the poor of the parish, but it is probable that it had a much earlier origin. It is further believed that it was held subject to the condition that on the 11th of November in each year the inhabitants should perambulate the Forest, and that, at the stroke of midnight, the oldest of them should drive an axe into one of the branches. Other parishes are said to have enjoyed similar privileges, but to have lost them through omitting this ceremony. But these beliefs are not supported by any evidence. Willingale's defiant act was treated by the Justices as a theft, and he, not having knowledge enough to assert his right before them, was convicted. His case was, however, supported by a member of the Commons Preservation Society, and the necessary funds having been raised, an action was commenced in support of the lopping right, and this step, involving a legal claim, resulted in an order by the Court for the maintenance of the *status quo* in Loughton Forest pending the decision of the case. Although this case dragged its slow length for several years, and was never finally decided, it had the most important consequences. The time gained, during which the Forest was practically saved from further inroads, allowed public interest in the question to grow, and brought into the field more powerful combatants, whose resources were used to raise wider and more far-reaching issues. Still the attitude of the lords of the manor was so confident, and the combined front which they showed was so firm, that they succeeded in persuading the Government of the strength of their case, and a Bill was actually introduced into Parliament by the First Commissioner of Works, which would have given the public only a paltry 600 acres, with leave to purchase at full market value 400 more, which were allotted as compensation to the commoners, the whole of the remaining 5000 acres being absolutely made over to the manorial owners. I well remember the meeting, which was held at this crisis, of the Commons Preservation Society, at that time the only organised body which seemed to care for the interests of the public. So dark was the outlook that it was seriously debated by the committee whether this so-called compromise should be accepted. It is not too much to say that the fate of the Forest trembled round the table where this committee sat. If weak-kneed counsels had then prevailed, the Bill would probably have passed without opposition, and the Forest, as we know it, would have ceased to exist. Happily a spirited policy carried the day. The Bill was vigorously opposed, and dropped, and Mr. Cowper-Temple subsequently carried against the Government an address to the Crown, calling upon it to preserve those parts of Epping Forest which had not been enclosed by legal authority. The immediate result of this was the passage in the next session of a Bill appointing a Royal Commission to inquire into the rights of the Forest, and prepare a scheme for its future management. In the meanwhile some isolated attempts were made to resist the enclosures. A resident removed the fence erected by the lord of the manor round Lord's Bushes, at the spot where it was crossed by a well-known ancient bridle-way. The lord of the manor brought an action in which he was unsuccessful, to the great joy and enthusiasm of the neighbours.

Public attention was now fully aroused, and a new body was formed, called the "Forest Fund Committee," who assisted in forcing the question to the front; but the public without a champion is powerless. Happily in the Corporation of the City of London a doughty one was at hand, able to contend even with the united forces of the manorial lords. It was Mr. Scott, the far-sighted City Chamberlain, who first suggested that the compulsory corn metage should be commuted into a small fixed duty, and applied to the preservation of open spaces; and the sinews of war having been thus provided, we owe it to Mr. T. C. Bedford—an able and fearless member of the Court of Common Council—that the duty of making a supreme effort for the rescue of the Forest was vigorously pressed upon the Corporation. Owing to the happy chance that the Corporation were owners of a cemetery at Wanstead, which gave them the right of grazing a cow or two, they were able to take up the cause of the public as commoners of Epping Forest. In August 1871 a suit was commenced against the lords of the manors, which lasted more than three years, and cost both sides not less than £30,000. The labour connected with it was enormous, owing to the multiplicity of interests involved and the obscure and intricate questions of ancient law which it raised. That it was conducted to a successful issue was largely due to the energy of the late City Solicitor, Sir T. Nelson, and the zeal, combined with knowledge, of Mr. now Sir Robert Hunter, who, as honorary solicitor to the Commons Preservation Society, had for several years made the law affecting common rights his especial study. It was contended on behalf of the lords of manors that each manor was separate, and that its commoners had no rights over the remainder of the Forest, so that, in fact, if he could succeed in satisfying his own commoners, each lord of a manor could enclose or do as he liked with it. On behalf of the Corporation, on the other hand, it was urged that there never were such boundaries, and that the commoners had always enjoyed the right of "intercommonage," as it was called—*i.e.* for their cattle to wander all over the Forest.

This tremendous legal battle, in which eighteen barristers were engaged, lasted for three years, and the final hearing for seventeen days. At length, on the 24th November 1874, the Master of the Rolls, Sir George Jessel, delivered a judgment which has never been surpassed for lucidity and conciseness, completely endorsing the contention of the Corporation, and for ever overthrowing enclosures, the illegality of which was left beyond dispute. So masterly and conclusive was the judgment that no attempt was ever made to appeal against it. The action was said to have cost the successful litigants £25,000. It is needless to say that this happy conclusion of the long controversy, which had cost so much anxiety and heart-burning, was received with general congratulation.

Soon after, the Commissioners, who for several years had been prosecuting their inquiries on the same lines as the Corporation suit, presented their report, which included a scheme for the future regulation of the Forest, but recommended very easy terms to the "grantees"—*i.e.* those who had bought land shown to be illegally enclosed. This did not meet with general assent, and Mr. Burney, a commoner of the Forest, took upon himself to bring down a large party of workmen, and demolished the fences of some of the grantees; and, although this action was condemned at the time, it had the effect of reminding the holders of this land of their precarious position under the decision of the Master of the Rolls if some scheme or compromise were not presently devised. Under stress of this dread, and under the skilful pilotage of Sir H. Selwin Ibbetson, M.P.,[1] the Act of 1878 was passed without opposition, and finally settled the whole question. It provided for the disafforestation of Epping Forest, which is henceforth to be under the charge of the Corporation of London as conservators. It is therefore no longer a Royal Forest, except that the Crown appoints the Ranger, in pursuance of which provision the late Queen appointed His Royal Highness the Duke of Connaught. The deer were ordered to be transferred to the conservators, and the rights of the commoners were preserved intact. It restored to the Forest the whole of the lands in the hands of the lords of the manors, and, with regard to those which had been purchased and on portions of which houses had been erected, it appointed an arbitrator, Sir Arthur Hobhouse,[2] to determine how much land should be allowed to be retained in each case as "curtilage" to the house, and what payment should be made for it. He had also to award compensation to the lords of the manors, the dispossessed grantees, and the possessors of lopping rights, now put an end to for ever, and to settle divers other questions. Many of these proved to be of great difficulty and delicacy, and it was only after the lapse of several years and the exercise of great industry and judgment, combined with tact, on the part of the arbitrator, that their intricacies were at length unravelled. The final result was that 5542 acres were preserved to the public.[3]

The management of the Forest was vested in a committee, called the "Epping Forest Committee," which has all the powers of the Conservators, and which consists of twelve members of

[1] Now Lord Rookwood. [2] Now Lord Hobhouse.

[3] The most important subsequent additions to the Forest have been—Oak Hill, Theydon, 11 acres; and part of Highams Park, 30 acres.

the Court of Common Council and four verderers, resident within the Forest, to be elected by the commoners. A superintendent has since been appointed, and a staff of keepers preserve order in their respective beats. The section of the Act which concerns us most, and which ought to be written in letters of gold, is that which provides that the Forest is to remain for ever "as an open space for recreation and enjoyment." This is probably the first time that the public right of the use of open spaces has been recognised in law. That it is fully appreciated is shown by the increasing numbers who annually visit what is theirs by inalienable right.

It remains only to say one word as to the future of this beautiful woodland. The Act provides that "the Conservators shall at all times as far as possible preserve the natural aspect of the Forest." Such a reminder was perhaps hardly needed. The body to whom this charge has been committed are fully impressed with the importance of providing, for those who live the artificial life of our great city, the means of studying nature where it is unrestrained by art. If there were any fear of their forgetting this, it is probable that they would hear of it from the Essex Field Club—a society which comprises some hundreds of members interested in natural history and archæology, and who bring to bear a vigorous public opinion in favour of the preservation of all the wild life which finds a home in the "waste." It is in its varied aspects that the greatest refreshment is to be found for the eye and the brain, weary of dead walls and the turmoil of streets. The general opinion, so unmistakably evinced, that the Forest shall remain a forest and not be civilised into a park, is but the expression of a true instinct. May the people of London, for all time continue to draw full draughts at this source, and to profit by the companionship and teaching of Nature.

EPPING FOREST

ITS TOPOGRAPHY

"I advise them to study large-scale maps."

THE MARQUIS OF SALISBURY.

THE routes which I have attempted to describe in the following pages lead through what I consider, after roaming through the Forest all my life, to be its most beautiful glades and thickets, but I by no means claim to have exhausted its charms. On the contrary, even with the intimate knowledge of every part which I have acquired, I almost daily light on some picture which I have never before observed. Still more will the discerning visitor be constantly tempted to diverge from the lines I have indicated, and discover fresh scenes for himself. That he will be amply repaid I do not doubt.

I have, as a rule, avoided the more formal roads, and sought the by-ways, where nature has had the freest play, and in these, owing to the absence of landmarks visible from the denser thickets, it is difficult to give clear and unmistakable directions. The numerous tracks which intersect one another are trodden out aimlessly by cattle, and are consequently more confusing than useful as definite guides. It is, therefore, probable that many will wander from the trail without intending to do so. To those who are not blessed with a bump of locality, and who fail to follow my descriptions, I commend a close attention to the maps, which are, I think, as clear and accurate as it is possible for the scale of 3 inches to a mile to be.

EPPING FOREST

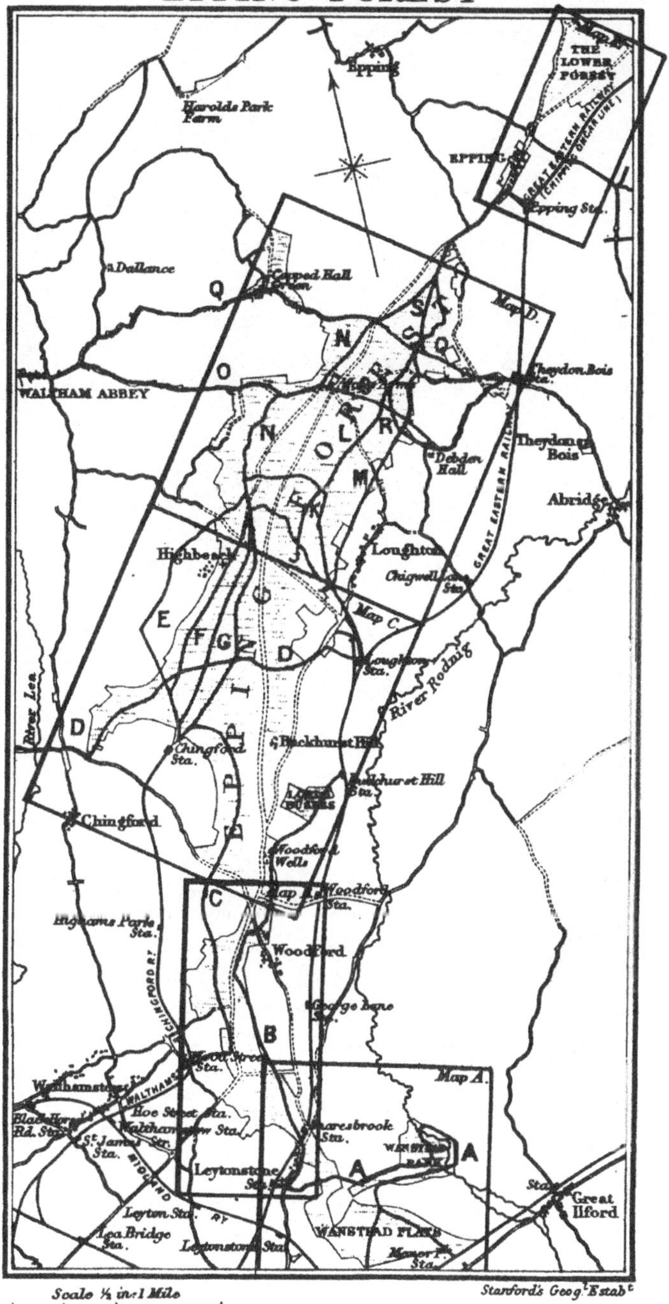

Scale ½ in.=1 Mile

1 ½ 0 1

Stanford's Geog. Estabᵗ

The intersecting lines, which divide them into squares of an inch and a half, are arranged in the direction of the four points of the compass. They are for the combined purpose of readily estimating the distance,—each side of a square representing exactly half a mile,—and of showing the direction, at a glance, to those who carry a compass, a practice which I strongly recommend.

The greater number of visitors to the Forest come by railway. I have, therefore, as a rule, started and terminated each route at stations, generally on the same line of railway.

It is obvious that the intelligent reader will be able to combine two or more of the routes to form a more extended expedition. Thus, after taking route G to High Beach, he can follow route K to Loughton, or route H to Epping.

The small figures on the maps show the height in feet above high-water mark. The routes described are indicated by the red lines. Hard gravelled roads are coloured yellow.

The small-scale index map opposite page 22 shows the position of all the railway stations relatively to the Forest. It also shows, for the benefit of those who drive, the chief roads of approach from London and Stratford.

It should be borne in mind that the days when east wind prevails are the best for selecting those routes that command distant views. Winds from the west or south come laden with London smoke which obliterates them.

ROUTE A

LEYTONSTONE TO WANSTEAD PARK AND FOREST GATE

5¼ miles

On leaving Leytonstone Station, cross the High Street, by the Red Lion Inn, and follow Harvey Road at right angles to it. This leads to the *Avenues* in 300 yards. These were planted 150 years ago by Sir Josiah Child, and are now, unfortunately, in a very decaying condition. They are but the remnant of a large number which radiated in several directions from old Wanstead House, and formed the approach to it. In the plan of Wanstead, as it was in 1745, they are well shown, extending as far as Snaresbrook pond.

Crossing the *Avenues* in an easterly direction to the road from Wanstead to Forest Gate, the entrance to the western arm of Wanstead Park is reached. [The same point may be reached by those driving or walking from Stratford by turning to the right out of the High Road, either along *Cann Hall Lane* or at *Harrow Green*, or by *Davis's Lane*. From Snaresbrook the park is reached by the new entrance road out of Red Bridge Lane, near Wanstead Old Church.] (For a history of the park, see p. 45.) Following the footpath through the *Reservoir* wood, the *Shoulder of Mutton* pond, the *Heronry* pond, and the *Perch* pond are passed in succession on the right, and the refreshment chalet on the left. Two hundred yards beyond the latter the footpath enters the beautiful woods which surround the *Great Lake*, and conducts to the old boat-house and the so-called grotto[1] above it, of which more need not be said than that it is a specimen of the style of landscape-gardening in fashion in the eighteenth century. The domed chamber is encrusted with shells and a mixed collection of stalactites, crystals, and looking-glasses. The best part of it is the view from the window. A glance at an old map shows that the river Roding formerly flowed through this lake, which was then at a lower level. At a subsequent date a course which now bounds the Park on that side was cut for it, to enable the level of the lake to be raised—an operation which had the effect of widening the channels and partly submerging the curious structure called the "Fortification." Previous to this alteration the straight cut canal-like branch of the lake, pointing east and west, gave a watery termination to the broad

[1] Burnt down, November 1884, but the façade remains.

Epping Forest

The sides of the Squares into which the Map is divided represent ½ mile.

and formal vista from Wanstead House. From a plan, dated 1735, in the possession of the Trustees to the Earls of Mornington, and prepared by a French landscape-gardener named Roche, who seems to have derived his inspiration from the gardens of Versailles, it appears that several other rectangular sheets of water of a similar character were designed, some of which could never have been completed.

From the grotto the path follows the verge of the lake [near the north end of the lake a path on the left leads to the high road to Snaresbrook Station] and, circling round its northern end, returns by the opposite side, passing the *Lincoln Island*, upon which is the heronry. Formerly the herons preferred the island in the upper pond, to which they gave its name, but many years ago they migrated to *Lincoln Island*, and have ever since built their nests in comparative seclusion on the tops of its highest trees. The crowds who now assist at the spectacle do not seem to have given them notice to quit. On the other hand I am rather alarmed at the great increase in the numbers of the rooks, which build close to them and which are continually chasing them in the air. To complete the circuit of the lake is a walk of rather more than a mile. To reach Manor Park or Forest Gate Stations, follow the southern boundary of the Park as far as the new road; this leads to Wanstead Flats, on the opposite side of which, in a southerly direction, is Forest Gate Station. If bound for Manor Park Station turn to the left on reaching the Flats, and opposite the gates of the Cemetery strike across the eastern corner of the Common.

ROUTE B

LEYTONSTONE TO WOODFORD OR BUCKHURST HILL

$4\frac{3}{4}$ miles to Woodford; $5\frac{3}{4}$ miles to Buckhurst Hill

After leaving Leytonstone Station, follow the High Street to the north for a quarter of a mile.

Immediately after passing the Workhouse School on the left, cross the railway, and enter the Leyton Flats—a level open space, much broken by old gravel pits, and covered with patches of gorse and broom. It is almost free from trees, and the few that are left have been seriously stunted and injured by the reckless gravel-digging of former years. Passing to the left of a small clump of trees, follow a broad green roadway which leads due north, and, leaving the Infant Orphan Asylum on the right, cross the Eagle Road at right angles. Leaving the Forest School on the left, *Gilbert's Slade*, a wide grassy glade, flanked by thickets of hornbeam and holly, and rather swampy in winter, conducts to the Woodford Road and Woodford Green. To reach Woodford Station, turn to the right, opposite the Castle Hotel. To continue the walk to Buckhurst Hill Station, follow the length of the Green, keeping to its eastern side, where a shady walk under chestnut trees is preferable to the High Road. From the north-eastern corner of the Green Monkham's Lane leads in 500 yards to an isolated block of forest known as *Lord's Bushes*, conspicuous for its picturesque oaks and beeches, and dense undergrowth of hollies. A well-defined track crosses this from the end of *Monkham's Lane* to Buckhurst Hill Station, at its north-east corner, but an hour may be well spent in exploring its beautiful glades. Another track from the same point crosses it diagonally in a northerly direction to the top of Buckhurst Hill.

ROUTE C

WOOD STREET STATION TO CHINGFORD

5½ miles

On leaving the station turn immediately to the left by *Upper Walthamstow Road* and follow the lane which leads up the hill to the eastwards and

The sides of the Squares into which the Map is divided represent ½ mile.

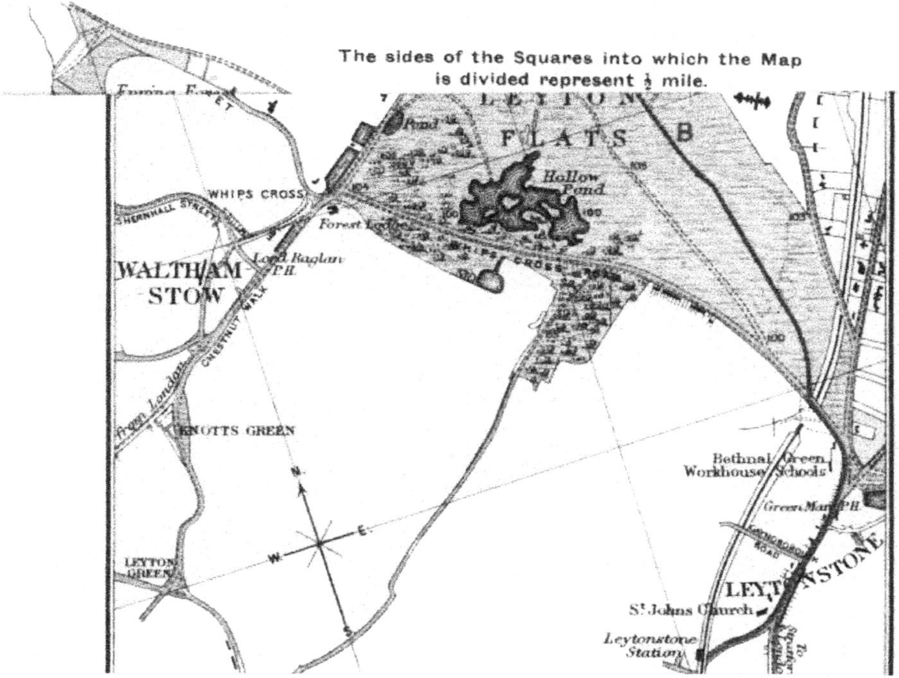

—

er
ya
re
F
tc
F
t

enters the Forest near St. Peter's Church, 400 yards from the station. This point may also be reached from Leytonstone by crossing the Leyton Flats as in Route B, but keeping somewhat more to the left, and crossing the Eagle Road at Forest Rise. From here follow the wide grass glade to the north as far as the Allotments. These, together with the land belonging to the East London Water Works, constitute a barrier across the route which necessitates a detour to the right as far as the high road. I have endeavoured, hitherto without effect, to persuade those who are responsible for these enclosures to allow the public right-of-way to be restored.

Regaining the line by *Hagger Lane*, from the corner of the Forest opposite the gate of the Allotments, a ride commences which may be identified by the distinguishing letter of this route marked on a pollard tree which stands by the opening to it. This leads over picturesque broken ground, and through a dense growth of holly and deciduous trees, in about ¾ mile, to Hale End Road. [From this point a pretty path along the eastern side of Higham Park leads to Woodford Green.] Crossing the road, enter the *Sale*, the green lane which skirts Higham Park, and thence Higham Park itself. It may here not be out of place to recount the steps by which this charming addition of 30 acres, comprising the lake of about 14 acres, and the best timbered part of the Park, was added to the Forest. A member of the Epping Forest Committee had for a considerable period sought to add this link which would connect the Walthamstow Forest with that portion which is called the Lops, and which was only attached to it by the above-named green lane. At length Mr. Courtenay Warner consented to part with it, for public purposes only, for the sum of £6000. A sum of £1500 was soon subscribed voluntarily. It now became of the greatest consequence to secure the adhesion and substantial contributions of the Local Boards representing the two parishes most interested. Those bodies took an enlightened view of their duty, but a legal difficulty intervened. While the law allowed Local Authorities to purchase land for recreation purposes, there was no provision under which they could contribute to the cost of land to be purchased, and owned by some outside body. In this dilemma the writer sought the advice of the Local Government Board. From Sir Hugh Owen the permanent under-secretary of that department, he received sympathetic aid. A Local Government "Omnibus" Bill was then passing through Parliament. It was then August, and the Bill had in fact reached the House of Lords and passed the second reading. It might be possible, even at that late stage, to introduce a clause legalising the action we proposed, provided we were not opposed. This was in fact done, and the clause passed both Houses without challenge. Thus new legislation was actually designed and carried within a period of ten days at the end of a session. This triumph was exclusively due to Sir Hugh Owen and the Rt. Hon. C. T. Ritchie, who piloted the Bill. It was followed by the generous contribution by the Corporation of London of the remaining moiety of the purchase money, and on the 5th Feb. 1891 the land became a corporate portion of Epping Forest.

Returning to the route, the gravel path conducts along the south side of the lake [the opposite side is an agreeable alternative] and finally emerges on the tops, the open grass slope which separates Woodford village from Chingford Hatch, and which is now the resort of scarlet-coated golfers.

The upper portion of the slope commands a striking view over the valley of the Lea. The stream of the Ching may be followed up on either side. Its right bank, *i.e.* its left looking up stream, is prettily wooded, but the other affords the driest and soundest walking after wet weather. Immediately after passing Whitehall Road Bridge, which must be crossed if one is on the right of the stream, leave the latter and pass close to the right of Forestside Cottages. Passing through the

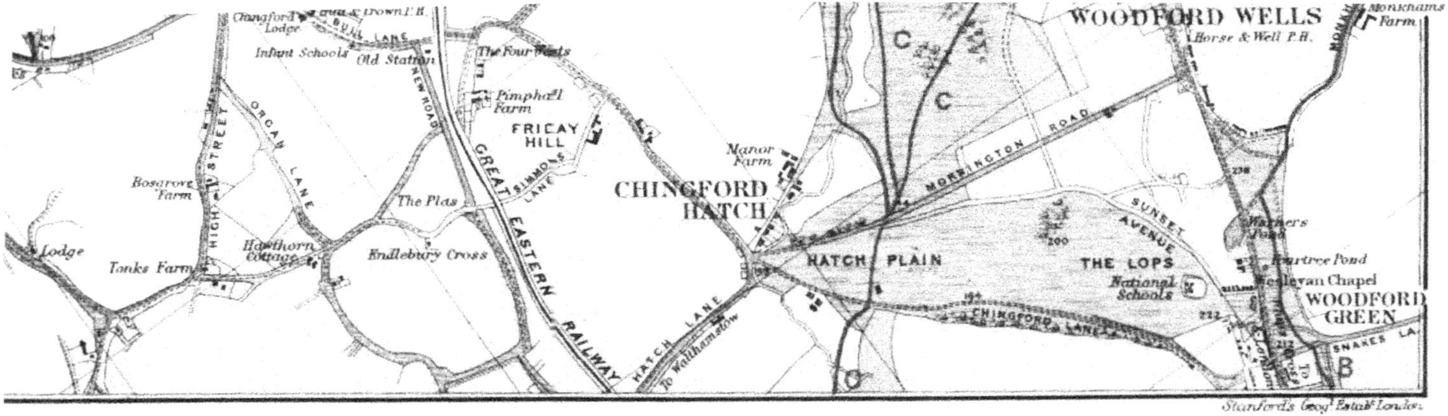
WOODFORD WELLS
Monkhams Farm
Horse & Well P.H.
CHINGFORD HATCH
Manor Farm
MORNINGTON ROAD
HATCH PLAIN
THE LOPS
SUNSET AVENUE
National Schools
CHINGFORD LANE
Wesleyan Chapel
WOODFORD GREEN
SNAKES LA.
Chingford Lodge
BULL LANE
Infant Schools
Old Station
NEW ROAD
The Four Wants
Pimphall Farm
FRICAY HILL
SIMMONS LANE
GREAT EASTERN RAILWAY
HATCH LANE
To Walthamstow
The Plas
Endlebury Cross
Hawthorn Cottage
ORGAN LANE
HIGH STREET
Bosgrove Farm
Tonks Farm
Lodge
200
Stanford's Geog! Estab! London.

grove of old pollard oaks, near Queen Elizabeth's lodge and the Forest Hotel, Chingford Station comes in sight. There are points in this walk which are decidedly swampy. I should be sorry, however, to discourage a trial of it, even in winter. In summer it is generally quite hard.

This route may be varied, by those who want to take the train at Buckhurst Hill, by diverging to the right soon after passing Chingford Hatch. Ascend the hill, and, crossing in succession the Whitehall Road and the Epping Road into the Loughton Road, follow this to the north for 100 yards, and turning to the right, immediately after passing Knighton Villas, strike diagonally across Lord's Bushes in an easterly direction, by a forest path which enters this piece of forest between two old pollard oaks at its south-west corner, and which leads to Buckhurst Hill Station.

ROUTE D

PONDER'S END STATION, ON THE GREAT EASTERN MAIN LINE, TO CHINGFORD OR LOUGHTON

2 miles to Chingford; 4¼ miles to Loughton

A road leads eastwards from the station across the meadows which border the River Lea. A mile from the station the Sewardstone Road, locally known as *Low Street*, is crossed at right angles. Four hundred yards beyond this, "*The Hawk's-mouth*," a narrow opening to the Forest on the left of the road, is reached. Turning in here across an open grass slope, the pedestrian ascends the hill which is crowned by the *Hawk Wood*, whence there is a beautiful view of the Lea Valley. Passing over the crest of the hill, and through the wood, he emerges at the edge of the wide open space known as *Chingford Plain*, also consecrated to the pursuit of golf, and from this point the Forest Hotel and Chingford Station, in the hollow to the right of it, are seen about half a mile distant.

The whole of the treeless space here seen was enclosed about 1860, and cultivated, but subsequently restored to the Forest. Though it is covered with grass, signs of its former cultivation are still seen in the ridges and furrows formed when it was arable land, and which sadly mar its grace. The Conservators will be grateful to any one who will suggest a cheap and effectual means of levelling these regular irregularities. To the left of this lies the great mass of the Forest, which extends from here nearly to Epping in a broad belt, 4 miles long, and comprises in one block about 4000 acres, nearly the whole of which is covered with a thicket, more or less dense, of oak, beech, hornbeam, crab, maple, thorn, birch, and holly. The whole of this wild and beautiful wood is worthy of exploration by the visitor, but from the absence of landmarks, and the intricacy of the glades and rides, it is not very easy to indicate clearly the best routes.

The immediate purpose being to cross it to Loughton, keep to the left and follow the highest part of the ridge, which leads in a curved line to the "*Bury Path*," an old forest way from Chingford to Sewardstone, which has recently been gravelled. Crossing this, immediately to the left of Mr. Earee's entertainment pavilion, seen on the right front, follow a narrow ride which leads from this point, in an easterly direction, to *Connaught Water;* or else, keeping rather more to the left, enter a picturesque green glade which runs eastwards from the "Woodman's Arms," a small inn on the edge of the Forest. This leads in a few hundred yards to a little winding rill called "*Cuckoo Brook*." Follow its right bank for 100 yards, then cross it by a well-defined track, and ascend the opposite hill to some shallow ponds known as the "*Cuckoo Pits*." Soon after passing these, the visitor, if he does not lose his way in the mazes of this part of the Forest, but keeps his head, and his easterly direction, comes in sight of *Grimston's Oak*, so named after the well-known patron of cricket, a fine tree standing alone in a circular clearing. Following the ride, at the end of which *Connaught Water* is seen, the *Ching Brook*, which flows into it, is reached. From this point the *Warren Hill*, across which the route lies, is seen in front. From the top of the hill there is a fine view of that part of the Forest which we have traversed, and as Loughton Station is visible from this point, further directions are needless. Except where the *Bury Path* is crossed, which is muddy after rain, I can conscientiously describe this walk as a dry one.

FAIRMEAD, FROM WARREN HILL.

FROM CHINGFORD TO HIGH BEACH

On quitting Chingford Station, the visitor at once enters the wide, treeless, grassy space, described in the last route, and sees beyond it the verge of the great thicket which extends from here, without a break, to High Beach, and, indeed, but for the intersecting roads, to Epping. This Fairmead Thicket is a fair sample of the denser parts of the Forest. The great majority of the trees (consisting chiefly of oak, hornbeam, and beech) are pollards, but among them there are a considerable number of straight-grown "spear" oaks, of 100 to 150 years' growth, which have never known the axe. In the drier parts of the Forest beeches to a great extent take the place of oaks. These "spear" trees will make fine timber for future generations, having received timely attention by being relieved of the competing growth of the unpicturesque hornbeam pollards. Consequent on these removals this thicket is becoming very dense in some parts, and owing to the confused maze of narrow cattle-tracks, it is not always easy to keep the true direction.

The walks to High Beach may be varied to any extent, but I will endeavour to indicate four routes which are tolerably well defined. It will assist the pedestrian to obtain a clear idea of the direction to follow if he will first visit Queen Elizabeth's Lodge, from the upper room of which he will obtain a good bird's-eye view of the district to be traversed. Let him take note of the spire of High Beach Church, which is the best landmark, though, from the inside of the thicket, it is only occasionally visible.

ROUTE E

3¼ miles

Cross the Forest by the *Bury Path*, which passes Mr. Earee's Retreat for the entertainment of children, visible from Chingford Station, and is continuous with the Sewardstone Green Road; follow the latter towards Sewardstone Green. About ten minutes' walk from the Woodman, where the road leaves the Forest, a very pretty lane on the left leads to Yardley Hill—a recent gift to the Forest by one of the Verderers. This hill, which projects into the valley of the Lea, commands a beautiful view of many miles of its course. From this point, by crossing the valley to Hawk Wood, a return to Chingford is readily made; or retracing steps to Sewardstone Green road follow it to Sewardstone Green, from the north-east corner of which a green lane leads by *Mott Street* to High Beach; or, better still, immediately after leaving the Green and passing Cashfield House on the right, turn to the right by the public footpath which crosses an open field and ascends the hill. Crossing *Leppets Hill Road*, you strike the path again a few yards to the right, by a barn, and from here it leads, in about a mile, to High Beach, close to the new church. Though this is not, strictly speaking, a Forest walk, the views from it over the Lea Valley are so beautiful as to justify me in inserting it here.

ROUTE F

2½ miles

Follow the *Bury Path* to a little beyond Earee's pavilion, then turn to the right along one of the several rides which run near to and parallel with the western verge of the Forest. From the point where *Leppets Hill Lane* enters the Forest there is a gravelled road along its edge northwards, though, if the weather is dry, this may be improved upon, by keeping a little more to the right, while pursuing the same direction, until High Beach is reached through the grove of old beeches. It is popularly supposed that these trees give the name to the place; this is an error, the true meaning of the word being the *High Beach* or *Bank*. This walk and those described in Routes G and H are apt to be swampy. The explorer should be well shod who undertakes them in winter. There is no need for this precaution in summer.

ROUTE G

2¾ miles

The corner of the thicket which approaches nearest to Chingford Station lies due north of it; a few yards to the right of the corner, a bell-mouthed opening shows the entrance to a ride, which leads in a tolerably straight course (a little to the east of north) to Fairmead Lodge, at the foot of High Beach Hill. About 500 yards from where the thicket is entered, the route leads at right angles into a green glade called the Woodman's Glade. Crossing this, the lower end of *Ludgate Plain*, which may be recognised by the picturesque half-timbered cottage at its upper extremity, is soon reached. Ascending the slope on the farther side of it, the path leads over a gentle rise and through another thicket to a more extensive plain known as the *Almshouse Plain*. From here the wooded slopes of High Beach Hill, crowned by the church spire, become visible in front. The next open plain that is crossed is called *Fairmead Bottom*, and gives its name to the picturesque Fairmead Lodge on the farther side of it—a well-known place of entertainment for parties of school children from London. On the farther side of the plain, and to the left of Fairmead Lodge, are two glades, either of which leads through the Beech Grove to High Beach Church.

CUCKOO PITS.

ROUTE H

2¾ miles

Keeping rather more to the right than in the last route, but to the left of the straight ditch below the Forest Hotel, enter the wood where a small stream, with a winding course, leaves it. Follow the left side of this for 300 or 400 yards, then cross it and ascend a low rise to the *Cuckoo Pits*. Immediately beyond these the *Green Ride* is entered. This must be followed to the point where it makes a sharp turn at right angles towards Fairmead Road. Instead of turning with it keep straight on to a wide opening in the thicket, visible in front, in the direction of the spire of High Beach Church. After crossing *Almshouse Plain*, ascend the hill to High Beach, leaving Fairmead Lodge on the right.

ROUTE I

3 miles

Those who prefer hard road or path to any of the above alternatives must follow the "Ranger's" road past the Forest Hotel as far as *Connaught Water*. Then turn to the left along its margin by the "red path." This leads into the road along *Fairmead Bottom* to High Beach.

ROUTE J

LOUGHTON TO HIGH BEACH

2¼ miles

On leaving Loughton Station, keep a straight course by the Station Road, through the village to Staples Pond. Enter the Forest on the left of the pond. A few yards from the head of the pond diverge from the wide green ride by a narrow ride on the left. This leads over a low rise to a rushy hollow known as *Debden Slade*, a corruption of Dead-man's Slade. Crossing the head of this, the track mounts the opposite hill diagonally to the old British camp and skirts round its south and east sides. From this point a group of tall elms at High Beach are visible and serve as a landmark. Continue along the crest of the hill in a north-westerly direction, so as to keep in sight the beautiful view to the south. From the point at which the Epping Road is crossed at right angles, a green roadway leads to the "King's Oak" at High Beach. This is one of the most beautiful walks in the Forest, but the following variation of it is even more charming, and will serve for the return journey, unless a more extended walk is desired, in which case return to Chingford by one of the routes already described, or to Epping or Theydon.

ROUTE K

HIGH BEACH TO LOUGHTON

3 miles

From a point on the High Beach road, about a mile north of the "King's Oak," strike due east across the thicket. This will lead to *Wake Valley*, sometimes called *Dick Turpin's* pond, an irregular sheet of water in a hollow by the Epping Road; cross the road and descend the valley through *Monk Wood*, keeping the left bank of the stream. Where this leaves *Monk Wood* and passes under the *Green Ride* [from this point a pretty track leads north-east up the hill to *Golding's Hill* ponds] cross the latter to some cleared spaces beyond. Continue to follow the left bank of the stream downwards till Baldwin's Hill pond is reached. From this point the driest and best track is on its right bank, until it flows in about half a mile into *Staples pond*, whence Route J started. Both these walks are fairly dry even in winter and after rainy weather.

tpath

ROUTE I.

BUCKHURST HILL OR LOUGHTON TO EPPING

From Buckhurst Hill, 7 miles; from Loughton, 5 miles

From Buckhurst Hill Station enter *Lord's Bushes* at its north-east corner, and, keeping near its northern boundary, leave it at the north-west corner. Follow the Loughton high road past Buckhurst Hill Church, and, 300 yards beyond this, diverge to the left across the cricket ground. A due northerly course will lead to Warren Hill. Leaving the Warren House, the residence of the Forest Superintendent, on the left, continue the northerly course across some cleared spaces, formerly cultivated—Strawberry Hill pond occupying the site of an old gravel pit is passed on the left—and by the green ride along the west side of Paul's nursery. This leads at right angles into the road from Loughton to High Beach, known as *Earl's Path.* A corresponding green ride on the other side of the road descends the hill to *Debden Slade*, a rushy valley, the head of which must be crossed. Ascend the opposite hill to the *British Camp*, described on p. 47. (To take this walk from Loughton, follow the directions given in Route J as far as the camp.) From this point, instead of following the hang of the hill, as in Route J, cross the heathery plateau at the top of it to the high trees of *Monk Wood*. A north-easterly direction will lead the visitor the whole length of the wood. This is considered by many to be the choicest bit of the Forest. Nowhere is there so wide an extent of well-grown trees, alternating with wild rushy plains, and intersected by deep picturesque dells. Cross the Loughton and Epping Road by the keeper's cottage, and descend the hill to the bottom of the next hollow, called "*Hangboy*" *Slade*. Follow this up the stream. The valley runs north at first and then bends round slightly to the north-east. Where the open glade appears to come to an end, turn to the right through the grove of beech-trees and up a little mossy hollow, until the Theydon Road is reached. On the farther side of this is an open heathery plain nearly bare of trees, owing to forest fires, at least this was the case until recently, but now countless young birches give it a pale green cast in the summer. After crossing the road, keep along near the thicket on the right-hand side. The tall firs on the left show the position of *Copped Hall Wood.* Our route lies at right angles to this direction. At the northern apex of the plain a well-defined track is found. This keeps along the side of the hill which slopes down to the left to a little stream, on the opposite side of which stand the fine beeches of "*Epping Thicks*"—probably the best grown trees in the Forest. At the point at which some villas on Piercing Hill come fully into view, 300 yards ahead, turn to the left, down a little hollow, and crossing the intervening stream into the "*Thicks*," rise the opposite slope to the wide *Green Ride.* Follow this ride to Epping.

I am aware that my directions for this walk are somewhat vague. From the absence of landmarks this is inevitable. It is of course needless to adhere rigidly to the actual route I have attempted to describe, which is susceptible of a number of variations, but if a general north-easterly direction is preserved from the old *British Camp*, the pedestrian may get bogged or hung up in a thicket, but he cannot go far wrong. It is at any rate well worth trying, for I should be inclined to select it as the most varied and attractive walk of any in the Forest. It is dry in all weathers throughout nearly the whole of its course.

ROUTE M

LOUGHTON TO THEYDON BY BALDWIN'S HILL AND OAK HILL

5¼ miles

Follow the ridge of *Baldwin's Hill* as far as *Golding's Hill* ponds, as described in Route P, or else keep along the bottom of the valley below it, hugging the bank of the stream to the point where it crosses the green ride, and leaves the high trees of *Monk Wood*. From here a track leads up the hill on the right in a direction a little north of east to the above-named ponds. At *Golding's Hill* ponds cross the Loughton road and take the green road along the eastern boundary of the Forest. After descending the hill, this road leads to *Debden Green*, but without leaving the Forest at this point keep a north-easterly direction and follow a green ride along the outskirts of the Forest, afterwards crossing Theydon High Wood to Oak Hill. This part of the Forest is a favourite resort of the deer. On reaching the Theydon Road, turn to the right and the station is reached in about a mile. I do not recommend this walk after wet weather. In summer it is hard.

ROUTE N

HIGH BEACH TO EPPING BY THE VERDERER'S PATH AND AMBRESBURY BANKS

3¾ miles to the Station

From the "King's Oak," at High Beach, follow the high road to the north for half a mile to the point where Honey Lane descends the Hill. Here diverge to the left to the edge of the hill, where a large mound has been thrown up to catch stray bullets fired on the rifle range which formerly existed here. From this point a ride follows the crest of the hill in a serpentine course, for about three-quarters of a mile, until it enters the road by Woodredden Hill. Follow the road to the right for 100 yards and then turn to the left into the Forest again, here known as *St. Thomas's quarters*, just before reaching the keepers' cottages, which stand on the edge of a small open plain. At the farther end of the plain is the opening of a narrow ride which must be followed. After passing on the left the Forest pound, an enclosure of several acres, surrounded by an unsightly six-foot iron fence, which is intended to serve the purpose of a "field pound," but is rarely or never used, the track strikes across the thicket to the north-east, and steers for the group of tall fir-trees in *Copped Hall*, a landmark which is occasionally visible. Crossing the road which leads to Copped Hall Lodge, the track conducts into the Epping road nearly opposite ***Ambresbury Banks***, an ancient British Camp described in the following chapter. Its position is indicated by the group of lofty beech and oak trees which have established themselves on the banks. Immediately to the left of the camp a beautiful glade, arched by fine beeches, leads diagonally from the high road into the broad "*green ride*" which is parallel to it, and which may be followed to Epping. This walk may be recommended in wet weather, as it leads over high ground and gravelly soil.

ROUTE O

THEYDON TO WALTHAM BY DEBDEN GREEN

4¼ miles to the Abbey; 5¼ miles to the Station

On leaving the station, turn to the left along the avenue which crosses *Theydon Green*. Thence the road leads to *Debden Green* in about a mile. A few yards before reaching the latter, at a gate on the right, a notice-board indicates that there is a right-of-way across the field to the Forest. From the point where the latter is entered a broad glade, overgrown with brake fern, ascends the hill. Looking back to the south over Loughton, there is a fine view of the Thames Valley to Kent. At the top of the hill there is a wide plateau, with scattered pollard oaks growing on it, but in summer covered with a dense growth of brake fern. Cross this in a westerly direction to the old roadway known as "*The Ditches*," which bounds the plateau on the west. On the farther side of this roadway look for a gap in the thicket. This is the top of *Copley Plain*—a grassy hollow which leads down into the Valley of *Hangboy Slade*, and, continuous with it, a ride ascends the opposite slope to the Wake Arms on the London and Epping road. *Dulsmead Hollow*, which falls into *Hangboy Slade* from the west, 200 yards farther to the left, may be followed instead of this ride, and leads into the Loughton road, close to its junction with the London road at the Wake Arms. From here take the road by *Woodredden Hill* to *Honey Lane*, or else follow for 500 yards the *High Beach* road, which also branches from the *Wake Arms*, and then turn to the right through the thicket. In a few yards the crest of the hill is reached, which overlooks the valley of the Lea. Below, at the foot of the hill, lies the green opening of *Honey Lane Plain*. Descend the hill through the thicket, which is rather dense in this part; and, at the corner of the green, by the covered drinking-trough, take the road to Waltham. With the exception of one or two spots this walk also is generally dry.

ROUTE P

LOUGHTON TO HIGH BEACH BY BALDWIN'S HILL

5¼ miles to complete the circuit to Loughton

To any one who desires a very dry walk through the best part of the Forest, I can recommend the following circuit by High Beach. It follows the road nearly the whole way, but keeps the highest ground, and commands charming views of Loughton Manor, the distant Kent Hills, and the Lea Valley. On leaving Loughton Station, pass through the village to Staple's Pond; here turn to the right along the road for 200 yards. By the pound ascend the grass slope on the left to a group of fir-trees at the top. From here a lane leads along the crest of *Baldwin's Hill* for half a mile, and commands an extensive view of the densely wooded Loughton Manor. This would be still more interesting if it were not for the unsparing pollarding to which nearly the whole of this otherwise beautiful wood was formerly subjected, and which gives it the effect of having been mown with a scythe. *Monk Wood* is a conspicuous object. Its billowing mass of tall trees rises like an island above the rest. Some sections of the slope of Baldwin's Hill, in the immediate foreground, were cleared of trees a few years ago, with a view to enclosure. A marked effect of this denudation is seen at one spot near the end of the ridge. The whole hillside, which is composed of viscous clay, no longer supported by the roots of trees, has commenced to slip down towards the stream below, producing cracks and fissures and other features in close imitation of an ice stream in the Alps. By *Golding's Hill Ponds* turn to the left along the Loughton and Epping roads, passing *Monk Wood* on the left. At the *Wake Arms* six roads converge. The one we have traversed leads from Loughton; the next in order, on the left, from Woodford and London; the third from High Beach; the fourth from Waltham Abbey; the fifth from Epping; and the sixth from Theydon. Follow the third to *High Beach;* or the fourth, for half a mile, as far as *Woodredden Hill*, and where the road begins to drop down the hill, turn to the left along the *Verderer's ride*, which also leads to High Beach. From High Beach either follow the road to the "*Robin Hood*," and thence by the road known as the *Earl's Path* to Loughton, or, immediately after passing Paul's Nursery, turn to the left by a beaten track, which is a short cut through a picturesque piece of beech-wood to the "Robin Hood," and from thence take the *Earl's Path*.

ROUTE Q

THEYDON STATION TO WALTHAM BY COPPED HALL GREEN

4 miles to the Abbey; 5 miles to the Station

Follow the road along the north side of the green, and, immediately after passing the church, turn to the right, in a N.N.W. direction, across the open grass field, where the remains of ridge and furrow show that it underwent a period of enclosure and cultivation before it was restored to the Forest. Where the thicket is entered, at the farthest corner of the field, a tiny stream leaves it. Follow this up, keeping it on the right. Whenever the valley divides into two branches, which it does twice, follow the fork to the right. When it comes to an end pursue the same direction, by compass, until, near some old gravel-pits known as "*Wormleyton Pits*," you emerge on a heather-clad space comparatively free from trees. From this point the *Copped Hall* group of Scotch firs, with a single columnar silver fir towering above them, to which I have referred in previous routes, is seen about half a mile in front. Taking this as a landmark, and steering as straight for it as the intricacies of the thicket will admit of, it will guide you past *Ambresbury Banks* to the point on the Epping road where the Waltham road leaves it. Follow the latter past the *Copped Hall* entrance gates, and *Copped Hall Green* to Waltham. This is a tolerably dry walk all the year round.

ROUTE R

THEYDON TO HANGBOY SLADE

4½ miles to complete the circuit to Theydon

From Theydon Station take the road to *Debden Green*. From its north-west corner a lane leads into the Forest. At the point where this is entered the small stream which flows from the valley of *Hangboy Slade* leaves it. If the course of this stream be followed upwards, keeping always near the bottom of the hollow, it will lead, in about a mile, to the Theydon road. Crossing this and the open heathery plain beyond it for 400 or 500 yards, turn to the right into the thicket, and, keeping a S.S.E. direction, *i.e.* parallel to the Theydon road, come out on to *Theydon Green* near the church. This very pretty circuit should be reserved for dry weather. It is easily identified by following the hollow upwards. I have only partially indicated it by a red line on the map, lest it should be confused with other routes which intersect it.

ROUTE S

THEYDON TO EPPING

2¾ miles to the Station

Follow the road past the church as directed in Route Q. Cross, in a N.N.W. direction, the green beyond it, as described in the last route. On entering the thicket, and after following up the right-hand side of the little watercourse for a few yards, turn to the right up the hill, whence a track to the N.N.E. leads to Piercing Hill, near some villas on the right. From here the tall red water tower at Epping comes in sight. Keeping a point or two to the left of this landmark, traverse the little stream at the bottom of the hollow, and pass through "*Epping Thicks*" to the town of Epping.

ROUTE T

THE LOWER FOREST

This very pretty wood, of about 300 acres in a triangular block, is isolated from the rest of the Forest, and lies just beyond the town of Epping. I shall not attempt to describe any routes through it, as they would be difficult to identify. But a summer's afternoon may be well devoted to its exploration; I say summer advisedly, as parts of it lie low and swampy. An ancient bridle-way, at one time the high road from Harlow to London, known as the "Stump Road," bisects it from north to south. *The Lower Forest* lies between Epping and North Weald Station, and equidistant from them. The prettiest road from the former station is that which passes through Coopersale; the shortest through the town of Epping.

The sides of the Squares into which the Map is divided represent ½ mile.

Epping Forest *Map E.*

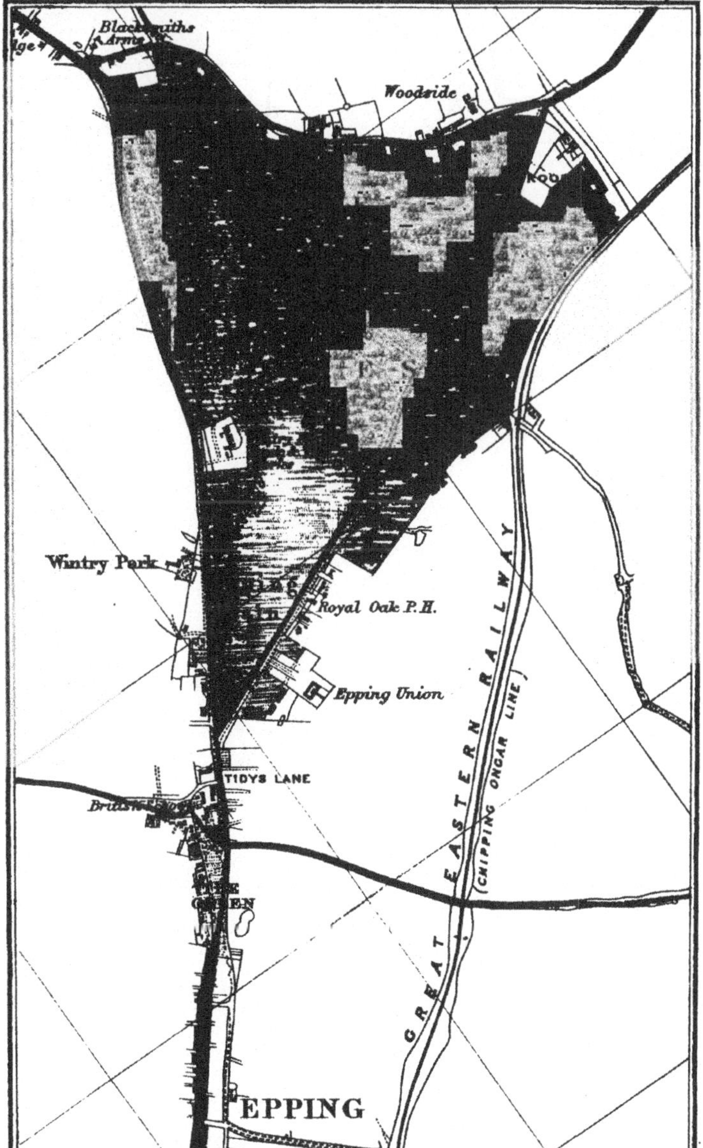

Stanford's Geog.l Estabt.

OBJECTS OF INTEREST WITHIN AND AROUND THE FOREST

"A monk ther was, a fair for the maisterie;
An out-ryder, that loved veneryе . . .
He gaf not of that text a pulled hen
That seith that hunters be noon holy men . . .
Greyhoundes he had as swift as fowel in flight:
Of pryking and of huntying for the hare
Was al his lust." *The Canterbury Tales.*

WANSTEAD PARK

Wanstead Park has a history exceeding in interest any other part of the Forest. Just above the round pond, called the Basin, is the site of the great mansion erected by the first Earl Tylney in 1715. Old maps of the neighbourhood show avenues radiating from the house in all directions, those to the north-west extending as far as the Eagle at Snaresbrook, and there appears to have been little distinction at that time between the Park and the surrounding Forest. The avenues were planted for the Earl's father, Sir Josiah Child, under the direction of the famous John Evelyn of Watton, the author of *Sylva*, who informs us "of Sir Josiah Child's prodigious cost in planting walnut trees about his seat, and making fish-ponds many miles in circuit in Epping Forest on a barren spot." The splendid mansion built by Sir Josiah's son in 1715 was the work of Colin Campbell, the celebrated architect, and author of *Vitruvius Britannicus*. Horace Walpole speaks of him and the place in a letter to Richard Bentley, dated 17th June 1755:—"I dined yesterday at Wanstead: many years have passed since I saw it. The disposition of the house and prospect are better than I expected, and very fine. The garden, which, they tell you, cost as much as the house, that is £100,000, is wretched; the furniture fine but without taste. . . . The present Earl is the most generous creature in the world. In the first chamber I entered he offered me four marble tables that lay in cases about the room. I compounded, after forty refusals of everything I commended, to bring away only a haunch of venison. I believe he has not had so cheap a visit a good while. I commend myself as I ought; for, to be sure, there were twenty ebony chairs and a couch and a table and a glass that would have tried the virtue of a philosopher of double my size." In 1775 Harrison visited the house and thus describes it:—"Before the front of the house is a long vista that reaches to the great road at Leighton Stone; and from the back front facing the gardens is an easy descent that leads to the terrace, and affords a most beautiful prospect of the river, which is formed into canals; and beyond it the walks and wildernesses extend to a great distance, rising up a hill, on the top of which the sight is lost by the woods, and the whole country, as far as the eye can reach, appears one continued garden. What a pity it is so fine an edifice, in so beautiful a situation, should be discarded by its possessor! and that a building calculated for the residence of the greatest subject in Britain should be inhabited only by a few servants."

The Earls Tylney were Hereditary Lords-Warden of the Forest, a dignity which descended to their successors in estate until the Epping Forest Act, 1878, deprived them of it. In 1784 the last Earl Tylney died, and Wanstead came to the Tylney Longs, whose heiress married the Hon. Wellesley Pole, immortalised by the well-known line in *Rejected Addresses*—

Long may Long Tylney Wellesley Long Pole live!

In the neighbourhood of Wanstead, however, Mr. Wellesley does not need *Rejected Addresses* to keep alive the memory of his extraordinary career. Such was his reckless prodigality that, having acquired by his marriage a rent-roll in Essex alone, raised under the influence of war prices to £70,000 per annum, he was within ten years of that time obliged to escape down the

Thames from his creditors in an open boat. His wife died broken-hearted, the custody of his children was taken from him by the Court of Chancery, Wanstead House was pulled down, and, though he had succeeded in the meantime to the headship of his own family as Earl of Mornington, he died a pensioner of his younger brother, the great Duke of Wellington. The estates came to his son, an estimable gentleman, but one who lived chiefly abroad and took no interest in them. Mortgaged up to the lips, the broad acres about Rochford, Ongar, Halstead, and elsewhere were sold. The remnant of the property the last Earl left away from his mother's family to his cousin on the father's side, the late Earl Cowley.

After being shut up for a great number of years, in 1880 the ornamental portions of the Park, comprising the woods, four lakes, and their islands, in all about 180 acres, were acquired by the Corporation of London in exchange for several pieces of Forest land, of great value for building, but of little or no use to the public, £8000 being paid in addition as a make-weight. The whole has been added to the Forest, but the grounds round the great lake are closed at night. There are three entrances to Wanstead Park—one from the south, near the Brickfield Pond, one from the west, towards the Avenues, and one from the north, by the new road from Red Bridge Lane.

AMBRESBURY BANKS AND LOUGHTON CAMPS

These two ancient camps are in the heart of the Forest, and owe their state of preservation in some measure to that fact, and their consequent immunity from the levelling action of agriculture. *Ambresbury Banks* lies about a mile south of the town of Epping, and close to the high road (see Route N); the latter on a buttress of the ridge to the north of *Earl's Path* (see Route P). Antiquarians have been much exercised about the origin and date of these earthworks; but their speculations have been to a great extent set at rest by the explorations conducted by the Essex Field Club in 1881 and 1882, under the able guidance of General Pitt-Rivers, and other experienced members of the club. Popular tradition attributes the origin of Ambresbury Banks to Queen Boadicea, and places the site of her final overthrow by Suetonius in this neighbourhood; but in the opinion of those best qualified to give one, there is no reliable evidence of this. On the other hand the verdict of General Pitt-Rivers is that this camp is undoubtedly of British origin, but whether erected before or after the Roman Conquest there is no sufficient evidence to show. The data upon which he founds this conclusion are, first, that the configuration of the ramparts is adapted to the features of the ground, instead of being constructed geometrically, which is the distinguishing feature of Roman camps; and secondly, the nature of the fragments of pottery implying a primitive condition of the art, and flint chips which were discovered in digging a section through the rampart and ditch. Those fragments which were discovered under the rampart itself, and upon the ancient surface-line, are necessarily of the same date as, or older than, the camp itself; and their character indicates a British origin. If the Romans had had any hand in the structure, it is all but certain that some remains indicating a higher civilisation would have been found in this part of the excavation. The height of the rampart appears to have been originally 10 feet, and the ditch of corresponding depth. Both height and depth are much diminished by the denudation of ages. Some think it unlikely that a camp would have been constructed in the midst of a wood, and it is inferred that the land must have been open at that time.

The *Loughton Camp*, which is distant about 2 miles south-east of *Ambresbury Banks*, is a smaller intrenchment containing 11 or 12 acres, but not so well defined; and is besides covered and partly concealed by a dense growth of beeches dwarfed by pollarding. There is even less symmetry of outline; and the constructors seem to have been guided solely by the desire to take advantage of every inequality for purposes of defence. To this the commanding nature of the site—a promontory projecting from a level ridge, and overlooking the marshy valley of *Debden Slade*, more than 100 feet below it—lends itself. Rabbits and foxes have taken advantage of the bank, and their operations have added to the destructive effects of time.

In the course of the investigation of the camp by the Essex Field Club, similar sections as in the case of Ambresbury were cut through the rampart and ditch, so as to expose the old surface-line, or the original floor of earth upon which the bank was piled. Here were found many flint "chips," and it is probable that these camp-makers used stone implements. Fragments of a primitive kind of hand-made pottery, pointing to an early British origin, were also found. The excavation showed that the section of the ditch in both camps was V-shaped, instead of the more common flat-bottomed form. General Pitt-Rivers says that the evidence before him is "sufficient to identify the camp as pre-Roman, and probably of a very early period."

The view from this point towards the south is one of the most extensive and beautiful in the Forest.

GREENSTED CHURCH.—Although five miles from the nearest part of the Forest, the interest which attaches to this church, both from an antiquarian point of view, and for the lover of the picturesque, perhaps justifies its mention here. Situated near Ongar, the terminus of the Woodford railway, it is best approached on foot from that town by a wide and straight grass avenue about a mile in length, which is terminated by the red-brick gables of Greensted Hall, close to which the church stands. The structural curiosity of the church lies in the walls of the nave, which are built of solid stems of oak-trees, or, as some say, of sweet chestnut, set upright, and tied together by tongues of wood let into perpendicular grooves at the edges of contact, a form of construction which, in the opinion of the learned, supports the belief that this part of the church dates from the Anglo-Saxon period, and renders it improbable that any part of it could have been taken out and renewed. The interior of the solid wooden wall thus formed has been brought to a plane surface by the adze, the marks of which are clearly traceable. Outside, the cylindrical boles of the trees are left rough, and their exposed surfaces are furrowed by the action of the weather into ridges and deep grooves in the direction of the grain, but are otherwise uninjured by the thousand years of exposure which they have probably endured, except that when the church was under repair in 1848 their lower ends were found to be partially decayed, and were cut off and replaced by a brick sill. The church is dedicated to the martyr-king St. Edmund, whose body rested here for one night, when, more than a hundred years after his death, it was carried to Bedrickesworth, now called Bury St. Edmunds, from London, whither it had been transported for safety in the stormy times before the Conquest, Greensted lying on the ancient road from London to Suffolk. The incident is thus referred to in an old chronicle of the Abbey of St. Edmund:—"His body was likewise entertained at Aungre [Ongar], where a wooden chapel erected to his memory remains to the present day." Edmund, who was more of a saint than a soldier, but had the courage to say to his conquerors, the Danes, who offered him his life if he embraced the religion of Odin and gave up half his kingdom, "You may destroy this frail body, but know the freedom of mind shall never bow before you," was bound to a tree and shot with arrows. It is related that, when the church was under repair in 1848, and the ancient timber lay on the ground, the old oak-tree near Eye, in Suffolk, to which tradition had always attached the scene of the martyrdom, fell to the ground, and, being cut up, an arrow-head was found embedded more than a foot in depth in the solid timber, and that the annual rings of growth showed that more than a thousand years had elapsed since it struck there.

It is a very pretty walk or drive of five miles from Greensted to Epping by Toot Hill and Ongar Park Wood.

OLD CHINGFORD CHURCH, which stands in a dismantled condition about a mile to the south of the present church on Chingford Green, is devoid of antiquarian interest, but is remarkable for its fine position, commanding a striking view over the valley of the Lea into Hertfordshire and Middlesex, and for the extraordinary trees of ivy, of great age and girth, which hide and protect the mouldering walls and roof. The spires and chimneys of Hackney and Walthamstow, seen dimly through a curtain of London smoke, are not an unpicturesque element in the view.

QUEEN ELIZABETH'S LODGE is a building of the Tudor period, constructed of massive timbers filled in with brickwork and plaster. It is popularly believed to have been the hunting resort of the Virgin Queen, and there is inherent probability in the tradition, though it is not confirmed by contemporary historians. All true foresters believe further that she was in the habit of riding up the staircase and dismounting at the top. Nor would this be a very difficult feat, as each step is a solid oak beam, and they are laid in short flights and at a moderate angle. On the first floor is another room of similar dimensions. This for many years was divided into bedrooms, but the partitions were removed in 1900 by the Conservators, to whom the Lodge belongs, and the room restored. These two rooms are now used to house various collections having a local interest. We owe these to the Essex Field Club, who take charge of them.

THE ABBEY OF THE HOLY CROSS OF WALTHAM, or rather the nave of it, is all that remains of the departed glories of the abbey, whose inmates exercised powerful sway over much of the adjoining country. The Forest formerly extended around it on the upland side, as the name, *Weald-ham* or Village in the Forest, shows. The River Lea, in very early times, formed a broad estuary on the other side, and it is related that, when the Danish ships lay at anchor here, while their owners were busy despoiling the country, King Alfred, by diverting the stream above or deepening the channel below, or by damming out the tide at Blackwall, left these ships high and dry, so that the Danes had to sacrifice them and save themselves by an overland flight. The spot originally

WALTHAM ABBEY.

derived its sacred character and its name of Waltham Holy Cross from a cross discovered by a holy man in Somersetshire, and thence miraculously transported to Waltham by oxen, acting under divine guidance. Tovi, standard-bearer to Canute, thereupon founded a church and religious establishment. King Harold greatly enlarged and enriched the foundation, and hither, tradition says, he came to pray before he went forth to meet the Normans, and hither his body was brought for burial after he was killed at the battle of Hastings. His tomb was within the chancel, inscribed with the words "Haroldus Infelix," but the destruction of this part of the building left it in the open air, and the stone which marked the spot was removed by depredators. It is said that in the reign of Queen

Elizabeth, a gardener digging came upon the stone coffin itself, which being opened the bones were discovered, but crumbled to dust on exposure to the air. The monks for many generations continued to exercise the powers and privileges granted to them, with varying fortunes, dependent on the favour, or the contrary, of kings, whose resort this monastery frequently was. Their influence was no doubt in many ways beneficent, and it is probable that we enjoy even at the present time some of the fruits of their arbitrary power. It was to the special permission to enclose, referred to on p. 111, and to the foresight often shown by the religious orders, that we probably owe the preservation as timber trees of the fine groves of beeches in the Forest, known as Monk Wood and High Beach Grove. Their relations with the surrounding people were not always of the most friendly character. Then, as now, rights of pasturage caused differences of opinion, though the following extract seems to show that they were not always so successfully championed as at the present day. Farmer, the historian, relates how, "when Simon de Seham was abbot, in the 30th Henry III. (1245), a dispute arose between the abbot and the townsmen of Waltham about the common lands. The men of Waltham came into the marsh, which the abbot and his convent formerly enjoyed as several to themselves, and killed four mares, worth forty shillings sterling at least, and drove away all the rest; the abbot was politely pleased for the present not to take notice thereof. Next year the same men of Waltham went to the abbot the Tuesday before Easter, in the name of the whole village, and demanded of him to remove his mares and colts out of the marsh. This the abbot refused to do, adding that if his bailiffs had placed his cattle otherwise than they ought, they might do well to have it amended, and yet so as to defer the matter till the Tuesday after Easter. On that Tuesday, Richard, brother to the King, Duke of Cornwall, came to Waltham, at which time both the men and the women of the town repaired to the gate of the abbey to receive the abbot's final answer.

"He put them off with the information that he was preparing for a journey into Lincolnshire to meet the justices itinerant, and said that he would settle the affair at his return. Not satisfied, they went into the pasture, and, in driving out the abbot's mares and colts, drowned three worth twenty shillings, spoiled ten more to the value of ten marks, and beat the keepers, who resisted them, even to the shedding of blood. Fearing, however, that they should be prosecuted on the return of the abbot, they desired a 'love day,' and offered to pay damages for the injuries committed; but, instead of doing so, they went to London and accused the abbot to the king of having wrongfully taken away their common land, and bringing up new customs, adding that he would 'eat them up to the bone.' The abbot then excommunicated the men of Waltham, and they impleaded him at common law for appropriating their common land to himself. They were unsuccessful, and after a long suit in the King's Bench, were glad to confess that they had done wrong, and they were amerced twenty marks, which the abbot remitted, and, on their submission, he *assoyled* them from the excommunication."

Waltham Abbey did not escape the heavy hand of Henry VIII., but, along with the rest of the monasterial foundations, the canons, abbots, and monks passed away, and their rich lands and forestal rights were surrendered to the king in 1540. The present abbey, which is a conspicuous feature from the high ground of the Forest overlooking the valley of the Lea, is but a fragment of the splendid old abbey, enriched and embellished by a long line of kings. As has been already shown, a sanctuary of some sort stood here from very early times. The abbey was completed in the 11th century, but many additions continued to be made to it by pious worshippers, until it was shorn of its glories at the time of the surrender.

What remains of it is said to be "the earliest undoubted specimen of the Norman style of architecture now existing in England."

Originally in the form of a cross, with a massive tower rising out of the intersection of the nave and transept, all but the western part of the nave was either destroyed or fell to the ground when left unsupported by the rest of the structure. The work of destruction must indeed have been carried out eagerly for us to find this extract from the churchwarden's accounts in 1556: "Anno 1556. *Imprimis* for coles to undermine a piece of the steeple which stood after the first fall, 2s." The present building has been carefully restored in recent years, and some of its ancient beauty again brought to light. The columns, each one differing from its neighbour, are well worthy of special study. The crypt, which Fuller, the historian of the abbey, and, for a short time, curator of the parish, described as "the fairest that ever I saw," the Lady chapel, recently restored by Sir T. F. Buxton, and the many curious monuments of worthies who have lived and died under the shadow of the abbey, are worthy of note.

EPPING FOREST

ITS ANIMALS

"The heath-stalker, the strong-horned stag—seeks shelter in the wood."—*Song of Beowulf.*

"The mole has made his run,
The hedgehog underneath the plantain bores,
The rabbit fondles his own harmless face,
The slow-worm creeps, and the thin weasel there
Follows the mouse, and all is open field."
Aylmer's Field.

We learn from an old writer that the beasts of the chase "were commonlie the bucke [fallow], the roe, the fox, and the marterne. But those of venerie in old time were the hart [red-deer], the hare, the bore and woolfe; but as this held not in the time of Canutus, so insteed of the woolfe the beare is now crept in, which is a beast commonlie hunted in the east countries, and fed upon as excellent venison, although with us I know not anie that feed thereon or care for it at all." Though some of the animals here mentioned have disappeared from these realms, there is still prolific wild life in these thickets which lends them an additional attraction. We owe the preservation of the Forest itself to the fact that it was formerly kept for the free range of beasts of chase, and it is fitting that we should pay them every attention. The greatest care is therefore taken to guard them from molestation. Guns are not allowed to be used. The result is seen in unusual fearlessness, especially of many of the birds, such as wild ducks. In the following memoranda on the animals and birds I have limited my observations to those characteristics which may be noted by any observant eye. I have not included in the list the prairie wolf, of which there used to be a specimen at the Zoological, which was alleged to have been caught in Epping Forest. According to the story which appeared in *Land and Water* in the summer of 1884, this animal was purchased as a cub by a gentleman living at Leytonstone, from a hay carter, who said that he had caught it

ROE-DEER. RED-DEER. FALLOW-DEER.

with two others in the Forest, and described them as fox cubs. When the cub grew up he turned out to be an undoubted coyote or prairie wolf. It was conjectured that their parents must have been turned out for fox cubs, and although the gentleman to whom their introduction was attributed has assured me that what he enlarged were undoubtedly Spanish foxes, it is just possible that some one else may have turned out coyotes, and that they lived and bred unsuspectedly for some years in the Forest. The story is, however, not sufficiently confirmed for me to include wolves among the Forest animals.

The **Fallow-Deer** are the most conspicuous and distinctive of the wild animals inhabiting the Forest. They have wandered there for many centuries, but are believed not to be indigenous, but to have been introduced by the Danes. No fossil remains are found, although those of roe and red-deer are frequently dug up. The deer, both red and fallow, as I have explained elsewhere, were formerly rigidly preserved for the use of the king, but some favoured individuals were allowed to hunt. For instance, Henry III. enacted, "Whatever archbishop, bishop, earl, or baron shall be passing through our forests, it shall be lawful for them to take one or two deer, by view of the forester, if he shall be present; if not, he shall cause a horn to be sounded lest it should seem a theft." Presents of venison were also frequently made in the following form:—"On sight hereof you are to kill and deliver to the bearer for the use of ——— one fat doe of this season, for which this shall be your sufficient warrant, and herein you are not to fail." At the beginning of the last century so many demands had been made upon the herd in this way, and by marauders locally known as "Waltham Blacks," that we find in the Court Rolls an order that "the stock of red and fallow being so low that they are likely to be extirpated, no more are to be taken for three years." After this they again increased, and our grandfathers describe them as being visible in large herds between Woodford and Epping by the passengers in the numerous coaches which passed that way, bound for the eastern counties. An old inhabitant writes: "When a boy my father took me for a treat to London. It was a fine summer morning. We started at four o'clock. I well remember the open plains bordering the Forest swarmed with deer. I am sure there must have been hundreds in sight at one time. This was in the year 1829." Another old inhabitant tells me that he remembers seeing them opposite Assembly Row, near Leytonstone. From encroachments on the area of the Forest, the depredations of poachers, and from the sporting rights of the Crown having been allowed to fall into disuse, the number of the deer diminished greatly from the beginning of the century; until, about the year 1860, there were said to be under ten left alive. Fortunately, before it was too late, attention was called to the subject, and owing to the fostering care of the Rev. J. W. Maitland, and subsequently of the Conservators, this unique herd was preserved, and has steadily increased, until there are, it is believed, about a hundred and seventy in the present year, 1897.

I call them "unique," as, though the same breed are found in some parts of Scotland, I believe these to be the only representatives in England of the ancient deer. The herds of tame fallow-deer, which are preserved in so many parks throughout England, differ completely from the Forest breed. The former are chiefly descended from ancestors imported from various parts of Europe, and, for the reason that peculiarities are preserved and transmitted, we find in these herds every shade of colour from white and dun to black.

The Epping Forest deer, on the contrary, show no such variations. They are all of a uniform dark brown, which appears to be black, except when one is in very close proximity. The mottles and spots which are so conspicuous on lighter-coloured fallow-deer, like those from Southern Europe, are invisible, except to a minute inspection of the skin after death. They are inferior in size to the park deer, and the horns are less branched, but whether this is due, as some think, to in-and-in breeding, or to the superior grazing enjoyed for many generations by the latter, I am unable to say. Their natural shyness is favoured by the density of the cover, which, from the difficulty of penetrating it noiselessly, gives warning of an approaching intruder. So well aware are they when they are concealed, that they will often allow one to pass within a few feet of their hiding-place without stirring. For this reason they are not always seen, and I have known many residents in the habit of traversing their haunts, who have even doubted their existence. There are always plenty within a mile radius of the Wake Arms, and the tracks of some of the larger bucks can generally be found in soft places within a short distance of Chingford Station. South of that point they do not now go. The visitor who wishes to get a glimpse of them may generally succeed in doing so, especially when the trees are leafless, by traversing the Forest noiselessly, at dusk, and up wind, in the neighbourhood of Loughton Camp, Monk Wood, St. Thomas's Quarters, Honey Lane, or the Theydon Thicket.

It may be said that in olden times the Forest was preserved for the sake of the deer, for the king to take his pleasure in hunting, and that without them it would probably have ceased to exist centuries ago. As I have explained elsewhere, the Forest laws, which were of unexampled severity, were mainly directed to this end. Although the deer were thus reserved for the king, there were some exceptions, and fee deer were allowed to certain persons, as the following very interesting fragment of an ancient Royal Roll, preserved in the British Museum, shows. The date is uncertain, but it is at any rate anterior to the dissolution of the Monasteries. The first paragraph is rendered unintelligible by mutilation:—

"*Item.* That the Lieu tenant, Rydyng forester, ye Ranger of the same Forest, certyfy at . . . . particulerly in a byll, the certentie of the deer kyllyd and servyd by every of them with the . . . to them directyd and gyvyn.

"*It.* The clerke of Swanymote every yere within xii days next after the fest of Saynt Michell, and within convenyent tyme after to make relacion to the kyng's hyghnes of the certentie of the deer kyllyd in the same forest in the sayd yere and byfore him presentyd as ys aforsayd.

"*It.* Yf any deer be kyllyd by chaunce and recovered so that the ffleshe be of any goodnes, then the keper in whose walke any such deer be recovered dilyver and bryng the same to the Lieu tenant, in his absens, to the

Rydyng forester and Ranger to th'entent that yf the sayd deer be mete for the Kyng's Hyghnesse then that yt be sent to his hyghnes, and yf convenyent tyme serve not, then that it be distrybutyd by the discretion of the lieu tenant, and in his absens by the sayd Rydyng forester or Ranger, best for the savegarde of the kyng's game.

"*It.* The sayd Lieu tenant Rydyng forester or Ranger to certyfy at the next Swanymote to the Clerke and withal to what persons the same deer so kyllyd by chaunce and recovered is distrybuted.

"*It.* That noo keper hunt or suffer any person to hunt within his walke for any fee deer but suche as before this tyme hath used to hunt for their disport and plesure for the kyllyng of their fee deer.

"*It.* Every keper yf he lyst to call his neyburs or borderers to assyst hym to hunt, and serve his warrant or commandment to th'entent they may take sport, and therby to increase frendschyppe and amytie.

"*It.* The freeholders to hunt and take disport for the kyllynge of the fee deer by the only syght of the lieu tenant, and in his absens, of the Rydyng forester or Ranger.

"*It.* That noo keper, his servants or deputies use or kepe in ther howses or walke any crossbow except yt be the lieu tenant.

"*It.* That noo keper suffer noo crossbow to be kept or used in his walke as nere as he can, except the person that so hath or use yt may so justyfy the keping therof by the kynges lawes, and yf he find or sees any suche crossebow, lett yt be dylyvred to the Clerke of the Swanymote to th' entent that dew punyschement may be . . . accordynge to the kyngs lawes and therupon informacion to be gyvyn in th' exchequer.

"*It.* Yf any keper suspect any person in his walke to kepe any crosebow in hys howse, then lett hym take with hym a verderer, or constable of the parische where the party so suspect dwellyth, and serche the howse. Yf any be founde, lett yt be dylivered as ys aforsayd.

"*It.* That every noble man may hunt and take his disport and plesure in the same forest [accordynge] at ther peryll accordyng as the kyngs lawes will suffer them, and the keper to gyve them the lokyng over, and to make them suche sport as it apperteynyth to a noble man, and the sayd keper to certyfye the same at the next Swanymote.

"*It.* That the warrants commandements and other fee deer be taken equally when yt may be moste conveyently sparyd by the outsyght and discrecion of the sayd lieu tenant Rydynge forester and Ranger, or one of them, in absens of the resydew.

"*It.* Noo keper or forester, lieu tenant Rydyng forester, Ranger or ther seruants hunt or any wyse by nyght, onless yt be for the recoverie of sum hurt or chased deer.

"*It.* That noo person taken by the kepers or any other offycers be commytted to warde or letten to bayll onles he be examyned by the lieu tenant, Rydynge forester, Ranger, and Steward or two of them.

"*It.* That the obligacions for the bayll remayn with the Steward of record.

"*It.* That every keper observe ther articles upon peyn of the forfetur of his offyce, and that upon dew examynacion and dew proof hadde byfore the lieu tenant, Rydyng forester, Ranger and Steward of the Swanymote, the sayd offender to be dischargyd of the exersyng of his sayd office, tyll the kyngs plesure be further knowyng.

"*It.* Thees fee deer to be alowed as here after folowyth:—

"ffyrst, the Justices of the Forest.
"*It.* The Lieu tenant.
"*It.* The Steward.
"*It.* The Rydyng Forester.
"*It.* The Ranger.
"*It.* Every Keper.
"*It.* My Lorde of Waltham.
"*It.* My Lady of Barkyng.
"*It.* The ffreholders."

Red-Deer—the largest and handsomest of the deer tribe in Great Britain—had free range of Epping Forest (along with the fallow) until the early part of the 19th century. The lord of the manor of Loughton has some heads taken from the Forest, which have been in the possession of his family for several generations, and which adorned what would certainly be accounted grand stags in Scotland at the present time. This was to be expected, as woodland red-deer always attain the largest size. From a statement made in 1871 by a very old inhabitant, and written down from his lips, it appears that the herd was of considerable size within his recollection. "There was a kennel at Loughton Bridge, kept by a man named Dean, and the dogs and horses for the hunt were kept there. A paddock was enclosed with high palings, and in this enclosure the deer, caught wild from the Forest, were kept, and hunted as required, but not killed. The black-deer (fallow) were hunted two or three times a week, killed, and used as required also. The method of taking the red-deer was by fixing a net (kept at Loughton Bridge for the purpose, and about a mile long) from the milestone in the Forest down to Monk Wood. The deer were driven up from Monk Wood into it. Two or three were then selected, and the rest, with the young ones set at liberty. A cart for the purpose was in readiness, and the deer caught were placed in it and taken to Loughton Bridge, where they were kept in the paddock till wanted." The following graphic account, taken from *Bailey's Magazine* a few years ago, is from the pen of an old writer well acquainted with the Forest:—"The limits of the grand old Forest have been grievously curtailed since the days when Mr. Long Pole Wellesley played high jinks at Wanstead House, where he kept a pack of staghounds, in a style of princely magnificence, to hunt the wild red-deer. These hounds were not foxhounds entered to deer, but the old-fashioned staghounds, such as King George III. used for the purpose. The servants were dressed in Lincoln green. There were constant hunt breakfasts at the Eagle, at Snaresbrook (then in the midst of the open waste), where all were bidden at Mr. Long Wellesley's expense. Everything was done with the most reckless extravagance; and he would scatter sovereigns to countrymen in the hunting field as readily as other liberal sportsmen would give shillings or sixpences. The pace was too great to last; and when the establishment at Wanstead was broken up, Tommy Rounding managed to secure a few couples of the hounds, which he kept in a rough sort of a way at the back of his house, the Horse and Groom, at Woodford Wells. It may seem passing strange to the present generation that a publican, living within nine miles of the London stones, should have kept hounds to hunt the wild red-deer; but there are those living who can vouch for the fact. Rounding was a capital sportsman, and so were all the family; his brother Richard, who had died previously, and his brother Robert, who only died last year. We must borrow from Mr. Thomas Hood, who knew him well, a description of the man himself:—

A snow-white head, a merry eye,
  A cheek of jolly blush;
A claret tint laid on by health
  With Master Reynard's brush!

And so the game was kept alive until an order came that the red-deer were to be caught up and taken to Windsor Park. This was carried out as far as practicable; the few that had escaped the toils of the yeomen prickers gradually fell victims to poachers and pot-hunters, until of the whole herd only one old stag remained. This stag was hunted by Tommy Rounding, and, after a great run, was killed at West Ham; and so ended the chase of the wild red-deer in Epping Forest. For years afterwards, at the festive gatherings at the Horse and Groom, a handsome silver cup used to be handed round, with the inscription—'From Long Wellesley abroad to Tommy Rounding at Home!'" The staghounds came to an end in 1797, but occasional hunts seem to have continued to be held, until early in the 19th century the red-deer were suffered, from the indifference of the authorities, to diminish; and the remainder of the herd were finally caught some twenty years later, and were transported to Windsor. A stag and two hinds were recently enlarged on the Forest in the hope of restoring the ancient breed, but they proved to be so mischievous that they had to be destroyed.

Roe-Deer. These beautiful little animals—the smallest of the deer tribe of Great Britain—roamed throughout the forest regions of Great Britain in primeval times, and doubtless Epping Forest formed no exception. Until quite recently, although they are very numerous in the dense woods of Scotland, they only existed in a wild state in one district in England—the wooded combes in the neighbourhood of the Vale of Blackmoor, in Dorsetshire, whither they were imported about the end of the 18th century. I had in February 1883 the satisfaction of reintroducing them to Epping Forest. Long convinced of the suitability of the Forest to their needs, I had at length the good fortune to secure the co-operation of Mr. Mansell Pleydell and Mr. C. Hambro, two Dorsetshire proprietors, in whose woods the roe are common, and who by every means in their power forwarded the experiment which I desired to carry out. Its success was also largely due to Mr. J. E. Harting, who took a naturalist's interest in the venture, and who not only superintended the capture, but accompanied them in their long journey through the night to Loughton. The capture was effected by long nets stretched across the wood, the deer being driven towards them by beaters and secured when entangled in the meshes, and carefully placed in a covered deer-van. In this way eight were taken, but two being destined for the Zoological Gardens, only six, two bucks and four does, were enlarged in their new home. Owing to the careful provisions which were made by Mr. Porter, an experienced deer-catcher, no hitch occurred, and when the door of the van was opened the next morning in the thickest part of the Forest, each one bounded out sound in limb, and after one stare of surprise, made off at its best pace into the cover. Subsequently I myself superintended a further capture and placed eight more in the Forest. They are now, in 1897, believed to number about twenty-five, but there are others which have taken up their quarters in coverts belonging to adjoining owners. They are frequently seen, especially in the neighbourhood of Monk Wood. The roe are less gregarious than most other kinds of deer, and, unlike the red and fallow-deer, which herd together in considerable numbers, they keep in pairs, or at most in small parties of three or four. They are of a bright reddish-brown in summer, changing to a dull gray in autumn. About a third of the size of the fallow-deer, their horns are not palmated and smooth as are those of that animal, but are branched and pointed like those of the red-deer, only on a smaller scale.

Hares are found throughout the northern part of the Forest, in moderate numbers. That well-known master of hounds, the late Mr. H. Vigne, had harriers in the neighbourhood for over sixty years, and none knew the Forest glades better than he. The pack is now hunted by Mr. Quare. With this exception they enjoy a quiet time of it as long as they remain in this sanctuary.

Rabbits have greatly multiplied in recent years. In fact there are now (1901) too many for the good of the Forest. They are now therefore in the limited category of "noxious" animals which the Conservators try to reduce, but it takes a good man to do it.

Badger. I understand that one of these animals was killed on the Hill Hall estate twenty years ago, and another more recently. They had not been seen for many years in the Forest until 1886, when I introduced several pairs, which have done well. They occupy a large "holt" in Loughton Manor, and are extending it every year. They draw out astonishing mounds of soil from the holes, and their underground township must be very roomy. They are tenants in common with the foxes. Indeed, it is an old fox-earth. Rabbits also occupy a portion. Though they generally live in harmony, I fear that the badger is occasionally guilty of the heinous crime of vulpicide. Both in 1896 and 1897 cub foxes were found mangled by them outside these earths. The badgers are not easy to see, being nocturnal in their habits, but the patient observer, who sits perfectly still outside their earths on a moonlight night, may be rewarded by a glimpse of them.

STOAT.

The **Fox**, the largest of our Forest beasts of prey, is welcome here to such pheasants and rabbits as he can catch. He does not often show himself, and the hounds do not trouble him more than once or twice in the season, nor does he yield them many trophies. They are often found sleeping in the daytime in the crown of some oak pollard. Several litters of cubs are annually produced in the deep earths near Monk Wood.

The **Polecat** was not uncommon in this district thirty years ago. As he preys on all other animals which he can catch, whether fur or feather, the gamekeeper preys upon him. He is consequently now rare if not extinct. As there is no game preservation, in the ordinary sense, in Epping Forest, but both the game and their natural enemies are equally welcomed, I hope that the polecat will return and multiply.

The **Marten**, which is closely allied to, but larger than the last-named animal, was till recently believed by some still to exist in the Forest. It frequents woods and lies in the hollows of old stems. Extremely agile in its movements, it traverses branches and leaps from bough to bough like the squirrel. "The last killed in Essex was trapped by Mr. Luffman, at that time keeper to Mr. Maitland, in April 1853, in one of the Loughton covers. I observed a living specimen near Ambresbury Banks on 29th July 1883."—J. S. E. The Rev. W. B. Daniel, in *Rural Sports*, published in 1801, says of this animal—"Some years since one used to run tame about the kitchen of the 'Bald Faced Stag' on Epping Forest."

The **Stoat** is the next in size to the polecat of the weasel family, and is of as sanguinary a disposition. Being the near relative of the arctic ermine it frequently turns partially white in winter. A pure white specimen has been killed in the Forest.

The **Weasel**, the smallest of his tribe, has all the ferocity and courage of his bigger brethren, and being even more active in proportion to his size, works havoc among the lesser birds and animals. I constantly see his tiny rope-like body arched like a bridge, as he races across my lawn, intent upon his bloodthirsty work.

Squirrel. Owing to the persecution which this popular favourite suffered, and the ease with which it was hunted to death when the trees had been recently pollarded and there were few tall branches in which to take refuge, it had at one time sadly diminished. Since the branches have been allowed to grow, this animal has had a chance to escape, and, as was to be expected, it has recovered the position which its beauty and pretty manners deserve. As acorns and beech-mast abound, it need not suffer for want of its winter store.

Otter. Two of these animals were killed by the Lea near Chingford in 1881. A few years ago one was seen to cross over the dry ground which separates the largest pond at Wanstead from the river Roding.

This year (1901) a pack of otter hounds hunted three or four days on the Lea and Roding and killed three otters. It is much to be hoped that all riparian owners will preserve these interesting fishermen.

Water Rats abound in the Ching, and are frequently found in the Forest ponds.

Dormouse. This pretty little rodent is easily tamed and made a pet of. In winter he lies in a state of profound torpor in the middle of a mass of grass, which appears to have been woven without a seam, and which must be torn open to find the owner, so that it is difficult to discover how he finished the fabric after getting inside. When found in this state he can only be awakened by baking him in front of the fire. He derives his name from this enviable capacity for sleep. I believe this to be the culprit who annually gnaws the heart of the young shoots of some of my chestnut trees, and litters the ground with them.

Long-tailed Field Mouse. A beautiful but mischievous little animal, who is fond of gnawing the bark of young trees and rooting up bulbs. In the Forest he is in clover among acorns and beech-mast.

Short-tailed Field Mouse. A larger mouse than the last named, though he resembles him in habit, and is equally destructive, especially to peas, beans, and crocus roots.

The **Hedgehog** passes most of his time in hiding, lying up in some dry ditch or hole. He never leaves his retreat by day, and, being then generally covered by the leaves and grass which adhere to his spines, he is seldom seen unless a dog discovers him by scent. The late wanderer in the Forest may, however, catch a glimpse of him in the twilight, quartering the ground like a well-trained setter, and seeking his food by the help of his nose. During winter he hybernates in a torpid state.

Water Shrew. "This elegant little animal is aquatic in its habits, frequenting clear fresh-water ditches and brooks. It swims and dives with great address. I have seen one swim across a pond in the Forest."—J. L. E.

Common Shrew. Is very gregarious. "I have seen at least a dozen, on a fine summer evening, gambolling together among the dead leaves when all is still. At the least movement they hide themselves in their burrows."—J. L. E.

REPTILES

The **Common Snake** and the **Viper** are found in the driest parts of the Forest. The former has yellow markings and a scaly appearance, and grows sometimes to three feet in length. The viper is of a dull brown throughout.

The **Slow Worm,** which is quite harmless, is not uncommon. It is of a bronzy colour, and about twelve inches long.

It is much to be hoped that these reptiles will not be exterminated. The first and last are perfectly harmless. The viper, though its bite would be inconvenient, and possibly dangerous, will not use its fangs unless attacked.

EPPING FOREST

ITS BIRDS

"The nightingale, of birds most choice,
To do her best shall strain her voice:
And to this bird, to make a set,
The mavis, merle, and robinet,
The lark, the linnet, and the thrush,
That make a choir in every bush."—DRAYTON.

MANY kinds of birds are observed in Epping Forest. Formerly they suffered much at the hands of bird-catchers, who caught great numbers with clap-nets and other devices. This is happily now prohibited by the Conservators' bye-laws, which are well enforced. Bird-catching is also now prohibited throughout the year under an order of the Home Office over a considerable area surrounding the Forest. The following notes are the result of the observations of several naturalists in the neighbourhood.

Peregrine Falcon. A pair of these birds were killed some years ago on the Copped Hall estate.

Sparrow Hawk. Frequently seen throughout the Forest. Is destructive to game and small birds. Though abhorred by the gamekeeper, I hope it will continue to be preserved on some of the estates adjoining the Forest. On the Forest itself it of course enjoys immunity from persecution.

Hobby. A pair of these beautiful but rare summer visitants nested for two years—1846-47. After being disturbed, they nested on the Hill Hall estate, and brought off their young.

Kestrel. Not uncommon. Seen hovering or gliding over the open ground. Feeds chiefly on mice and such small game. Nearly the whole of the proprietors round the Forest have agreed to protect this bird, which does little harm and much good.

Merlin. Reported to have been seen recently in Loughton Parish.

Common Buzzard. Has been seen several times during the last few years.

Rough-legged Buzzard. A pair of these birds killed some years ago at Epping were presented by the late Mr. Doubleday, of Epping, to the British Museum, where they are still preserved.

Honey Buzzard. A specimen of this bird was observed in September 1881 by my brother, Sir T. F. Buxton. It rose from a bush of beech close to his feet, and flew with rather a heavy flight to an oak about 20 yards off, where it remained for a few seconds. On examining the bush he found it swarming with wasps, which

SPARROW HAWK.

began to fly out on his touching the bushes. He then found pieces of wasps' comb lying on the ground outside the bush, and the dead leaves scattered around. On the following morning, on approaching the spot, he again saw the bird flying over the trees away from him. Much more of the wasps' comb was lying about, and a large cavity could be seen under the branches, where the nest had evidently been scratched up, and the contents strewn around. The branches of the bush showed no signs of injury, as they would have done had any man disturbed the nest. The spot was near the spring on the Woodredon Hill. The bird was subsequently seen on several successive days.

Raven. "A pair of ravens used to nest in the Wanstead heronry. Your grandfather, Samuel Gurney, took me there one spring (I think about 1833 or 1834), and we then saw two young ravens which the keeper had taken out of the nest a few hours previously. Your grandfather bought them, and turned them loose in the stable-yard at Ham House, but as he would not have their wings cut, in a few weeks' time they flew away."—J. H. G.

Barn Owl. I used to hear this bird nightly twenty years ago in Lord's Bushes, the old hollow trees of which it frequented. For several years it had disappeared from that locality, but I have recently heard it again. I saw a pair close to Fairmead Lodge in the summer of 1884. A chorus of angry jays attracted me to the tree where they were. His silent flight is caused by a fringe of down on the wing. Mr. Lister writes:— "Was a frequent visitor to my garden years ago, and used to beat over the Virginian Creeper that covered the end of the house, where sparrows roosted in great numbers. It bred in an old poplar in the grounds of Forest House, and still breeds yearly in the trees at Elmhurst, Woodford." This and the other species of owl are now preserved by the landowners surrounding the Forest.

Tawny Owl. Not uncommon; may be heard in the evening at almost any season. Monk Wood.

Long-eared Owl. Has been obtained several times, and it would multiply rapidly if it were not so frequently destroyed by gamekeepers.

Short-eared Owl. An autumn and winter visitor. Shot on several occasions by sportsmen in turnip fields in the neighbourhood.

Red-backed Shrike. One of the handsomest of our spring visitors. The broods keep together well into the summer, and are often seen. They breed in the neighbourhood frequently, if not every year. A pair built near Debden Green in 1895. "I have watched a male bird fix the head of a young wren on to a thorn in his 'larder,' using great exertion. Their store usually consists of insects and worms."—A. L.

Great Gray Shrike. The late Mr. Doubleday, of Epping, had a bird of this species which was captured by a bird-catcher in the Forest, and which he kept alive in a cage for a long time. It invariably hung up its food in its cage. If half a dozen small birds were put in (dead), it hung them all up by forcing their heads through the wires of the cage; and pieces of meat were also hung up.

Rook. The Forest is enlivened by several rookeries by Leyton Flats, Woodford Green, Copped Hall. They have vastly increased at Wanstead Park, where they breed on the islands.

Jackdaw. They come in great numbers to roost on the islands in Wanstead Park, where they also breed.

Carrion Crow. A few are always about; but they are becoming rare in most parts of England. They nest in several places in the neighbourhood. Their nests may always be distinguished from those of rooks by the lining of wool.

Hooded Crow. Occasional winter visitor.

Magpie. This predatory foe to other birds has become very uncommon hereabouts. I do not think there are at present more than two or three pairs in the whole neighbourhood, but they now receive special protection from the surrounding landowners. They used to breed regularly in the Rectory garden at Loughton. A pair used to breed at Knighton, but I have not observed them for many years. I know of two pairs breeding in the neighbourhood this year, 1897.

Jay. The character of the thicket has greatly encouraged this bird. The harsh rasping note with which he greets an intruder, and his sly ways as he flits ahead, always impelled by curiosity but always out of reach, enliven the Forest, and yet this is the only bird upon whose unlimited increase in the Forest I would place a check. He is cruelly destructive of all other birds' nests, except those which build in holes. For this reason a solemn order has gone forth from the Guildhall that the numbers

JAY.

are to be limited, and occasionally an effort is made to destroy some of them, with, as I think, some effect in increasing the numbers of lesser birds, and the liveliness of the thickets.

Starling. Perhaps the most abundant bird in the district after the sparrow, but he was a very rare bird in many parts of England fifty years ago. In hard weather about thirty come every morning to be fed on my lawn.

Ring Ousel. A fine cock-bird seen on Mr. Venables' wall and in his garden at the entrance to Wanstead Park on 5th September 1877. "Many years ago I saw one in my father's fields at Upton at the time of the spring migration."—A. L. One was seen in the spring of 1884 by the River Roding, and other instances have been recorded.

Song Thrush. Abundant in the autumn, but almost absent in mid-winter.

Blackbird. Abundant. We owe much of the music of the woods to these two birds.

Redwing. Large numbers roost in the denser thickets during hard winters, especially when there is a good supply of holly and other berries. At sunset they come trooping in from all quarters, and sweep round the trees in graceful flight before settling down for the night in the lower brushwood.

Fieldfare. In hard weather they come close to the houses to feed on holly berries.

Missel Thrush. Common. His grating note is heard in Wanstead Park in mid-winter and the early months.

Wryneck. Comes with the cuckoo, and is frequently heard in April, but soon leaves us for the more retired parts of the Forest.

Green Woodpecker. May be constantly heard tapping, and occasionally seen, in the northern part of the Forest. Very frequent about Cook's Folly, near Walthamstow.

GREEN WOODPECKER.

Greater Spotted Woodpecker. Fairly common in several parts. The rattling of his beak on a hollow branch is audible at a long distance.

Lesser Spotted Woodpecker. "Three times seen in my garden at Leytonstone."—A. L. Not very uncommon in the Forest.

Nuthatch. Generally found in the neighbourhood of old elm trees. In such positions, if supplied with nuts, they carry them off every morning, invariably leaving unsound ones. "Many years ago a pair used to come to my whistle every morning to be fed at an old mulberry tree in my garden. They would run down the branch to within a yard of my outstretched hand, but never quite ventured to take the nuts I offered them until I laid them down."—A. L.

Hoopoe. More than one example of this rare bird has been observed in the neighbourhood. One bird frequented a garden at Knotts Green for several days some years ago.

Tree Creeper. Resident and common in the Forest, and in Wanstead Park.

Cuckoo. A common summer visitor.

Kingfisher. Not unfrequently seen by the ponds and streams of the Forest. I have seen them flash out like a streak of blue lightning from the ditches close to the Forest Hotel at Chingford. They come to one of my ponds every spring.

Crossbills. The late Mr. Doubleday, of Epping, procured examples of both the common and European white-winged crossbills at Epping, and in September 1861 three specimens of the parrot crossbill were killed at one shot

KINGFISHER.

by a boy at Lamborne (see *The Zoologist*, 1861, p. 7759). A pair of the common crossbill nested in some firs at the Bower, close by Epping Railway Station.

House Sparrow. Everywhere.

Tree Sparrow. Common about the pollard willows near the River Roding.

Brambling. More frequent in the winter than is usually supposed. A few years since there were large flocks feeding upon the beech-nuts.

Greenfinch. Common. Large flocks may be seen assembling on the tops of trees, at dusk, in the winter.

Hawfinch. Common in the Forest, but very shy, and consequently rarely seen. The nests are so loosely built that the eggs may often be seen from below through the bottom. Mr. E. Barclay informs me that in 1878 a pair nested in Cook's Folly and brought off five young ones. "Occasionally seen in my garden in winter feeding on hawthorn berries. In June, about twenty years ago, I shot a young bird which was feeding on my peas in company with one or two more." An interesting note on the food and habits of the hawfinch, as observed at Epping by the late Mr. Doubleday, will be found in *The Zoologist* for 1843, p. 40, and 1856, p. 5098. He remarked that this bird is particularly fond of the seeds of the hornbeam, and is always more common in the Forest when a fine warm spring has favoured the

HAWFINCH.

flowering of the hornbeam and produced a plentiful crop of seeds.

Bullfinch. It used to be frequent throughout this neighbourhood, but it is so predatory among the fruit trees, that I am not surprised that it has become scarcer.

Chaffinch. Very common, and nearly as mischievous as the preceding.

Common Bunting. "Used to be more frequent than now. I have often had nests with eggs brought to me by the mowers."—J. L. E.

Snow-Bunting. One example of this bird was killed by a boy with a stone on Epping Plain. Another was killed in 1840 by an old sportsman who is still alive.

Yellow Hammer. Common.

Black-headed Bunting. Common in summer in bushes along the Roding. Occasionally seen in winter.

Goldfinch. Used often to be caught by bird-catchers on Wanstead Flats. Common about the lanes of Loughton.

Siskin. Occasional winter visitor to Wanstead Park. I saw one about the alders in 1883; three were seen in February 1884.

Linnet. Large numbers were taken by bird-catchers on

Wanstead Flats a few years ago. I have talked with these men when at work with their clap-nets, and they told me they often caught lesser redpolls; these and the linnets they put in cages; but greenfinches, which they took in prodigious numbers, were killed at once, and tied by the neck in bunches of a score each, and sold to the London markets. The numbers were far greater than could have been produced in the district, but in spring and autumn, when the bird-catchers ply their trade, there is a general shifting of quarters by these birds, so that many caught in the Forest must have come from a distance.

Lesser Redpoll. Loughton Forest and Wanstead Flats. Often seen feeding on the silver birches.

Mealy Redpoll. This pretty species only appears at long intervals, and, like the crossbills, in considerable numbers, probably in quest of food. "Some years ago many examples were trapped by myself and H. Doubleday."—J. L. E.

Skylark. Common and generally distributed.

Meadow Pipit. Only a winter visitant here. Its quickly repeated note, as it rises with jerking flight, cheers the sadness of our flats in winter.

Tree Pipit. Frequent about the borders of the woods. As soon as it arrives in April its song is constantly heard as it soars to a considerable height and descends with motionless and outstretched wings.

Common Wren. Everywhere.

Robin. Everywhere.

Pied Wagtail. A common resident. The greater number go south, to return about the end of February dressed in their summer plumage. Those that remain with us retain their winter garb until the spring is farther advanced.

Yellow Wagtail. Used to be frequent in summer, but now scarce in this neighbourhood.

Gray Wagtail. Not uncommon along the Roding in winter.

Bee-eater. Observed at Wanstead by H. I.

Spotted Flycatcher. A common summer visitor; but the latest of all to arrive except the nightjar.

Pied Fly-catcher. A female bird of this species was taken in the Forest by a bird-catcher on the 13th September 1877, and was forwarded to Mr. Harting for identification. It was kept alive for some days on chopped egg and meal worms. Two examples were shot in the Theydon Grove, and are in the Doubleday collection.

Nightingale. Well distributed over our district. The cock bird arrives first and sings best while he is waiting for his mate to join him. When the nesting is far advanced, his song is reduced to a tuneless croak. If a trap is baited with meal worms they do not seem able to resist the temptation, and they are consequently easily caught. In the spring of 1858 an old Leytonstone bird-catcher caught thirty-four about the avenues. I lately found one of these gentry, who mostly hail from Whitechapel, outside my fence at Knighton. He was at the time engaged in catching cock chaffinches with bird-lime, using a tame hen chaffinch as a decoy—an old device. In reply to the hope which I expressed that he did not catch nightingales, he replied, "No, sir, we seldom catches 'em, and when we does, *we lets 'em go.*" From the preternatural innocence with which this was said, it was perhaps true. The nests are placed on the ground in dense herbage, and are fortunately difficult to find.

Redstart. Frequent in summer, and a great ornament to our Forest.

Hedge Sparrow. Very common.

Alpine Accentor. Many years ago a specimen of this rare little bird was shot in a garden on the borders of

GREAT TIT.

the Forest by Mr. James Pamplin of Whip's Cross Nursery, Walthamstow. It was taken to London and identified by the late Mr. Gould, and will be found recorded in the *Magazine of Natural History* for 1832, p. 288.

Garden Warbler. Its pleasing song not unfrequently heard in April and May in forest copses and Wanstead Park; it used to be common about the avenues.

Sedge Warbler. Common in the spring by the Roding, at Red Bridge, and in Wanstead Park.

Grasshopper Warbler. Common throughout the district.

Blackcap. Arrives early in April, and is common in the forest and gardens, but especially in Wanstead Park, where the wood echoes in early summer with its fine song.

White-throat. In every hedge in spring and summer.

Lesser White-throat. A regular summer visitor to our gardens; a restless noisy little bird.

Chiff Chaff. Our earliest spring visitor.

Willow Wren. Very common everywhere when April comes. The approximate dates of arrival of the three species of Willow Wren in the Forest are—Chiff Chaff, 31st March; Willow Wren, 3rd April; Wood Wren, 15th April.

Wood Wren. Local rather than rare, generally frequenting tall trees, and the latest of the three willow wrens to appear, arriving usually in the third week of April. All three species breed in the Forest. The nest of the wood wren may always be known from those of the others by never having any feathers in the lining.

Golden-crested Wren. Common throughout the year.

Firecrested Wren. A bird of this species, taken by a bird-catcher at Epping, 26th November 1878, is preserved in the collection of British birds in the British Museum.

LONGTAILED TITS.

Wheatear. A passing visitor in spring and autumn, generally seen on Chingford golf links.

Whinchat. A summer visitor, breeding in the furze bushes in the open forest.

Stonechat. Breeds on the open forest; a few remain throughout the year.

Great Titmouse (Oxeye). Common all the year. A noisy pugnacious little bird, who resents, with angry chatter, intrusion into its thickets. They freely build in boxes placed for the purpose against the trees in my garden. I have more than once known the hen bird, when covering the young ones, to fly at my hand when I have opened the lid. They appear to be very fond of the seeds of the yew. "Oxeyes have built regularly for five-and-twenty years in the hollow of an old cherry-tree on my lawn, 3 feet from the ground. One season the usual time had passed without their beginning operations, and I found some stones had been dropped into the hole; these I removed, and the following day the nest was begun and the brood was brought off in due course."—A. L.

Blue Titmouse. Common all the year.

Cole Titmouse. Not uncommon in gardens, as well as in some parts of the Forest.

Marsh Titmouse. Common. Its name is misleading, as it is not a marsh bird in any sense.

Longtailed Titmouse. Resident and not uncommon. Families of ten or more keep together throughout the winter and flit from tree to tree, scouring them for insects. They never come to the lump of fat which is put outside my house for the other titmice.

Sand-Martin. Numbers used to breed in holes in the railway cutting between Leytonstone and Wanstead. They are the earliest of the swallows to appear, often arriving in March.

House-Martin. Not so common as they ought to be. The vast number of sparrows in the neighbourhood is hostile

WOODPIGEONS.

to their increase. A friend and neighbour increased the number of house-martins' nests on his house from two to ninety, simply by waging unsparing war on the sparrows.

Chimney Swallow. In the late summer these birds gather in thousands to roost in the reed-bed in the pond in Cook's Folly. This seems to be the general rendezvous for the district, and at sunset they may be seen flying in small companies from every direction towards this point. The confused noise as they swarm like bees to gain a foothold among the reeds is most remarkable. They also use a large clump of bamboo by my pond for the same purpose. In August 1855 a pure white swallow was observed at Epping.

Swift. Common from May to August.

Goatsucker or Nightjar. Frequently seen during summer in the Forest, where they breed on the ground, laying two oval eggs of a marbled gray colour. Unless my eyes deceived me I saw one, when suddenly disturbed, carry off one of its eggs between its thighs.

Woodpigeon. Great numbers breed in the Forest, and their exquisite note may be constantly heard. When there is much beech-mast immense flocks swarm in Monk Wood and Theydon, and the ground even looks gray with their droppings.

Stockdove. Always come in May and nest. It nests in holes in trees.

Turtledove. A constant summer visitor. I have seen flocks of them about Walthamstow in late summer. Its plaintive note may be frequently heard in Theydon Thickets at the time of incubation.

Pheasant. There are a fair number of wild-bred pheasants in the Forest, and many reared in the neighbouring woods come in for acorns in the autumn.

Partridge. A good many pairs, both of the gray and red-legged kinds, breed on the more open parts of the Forest.

LAPWINGS.

Thick-knee Stone Curlew or Norfolk Plover. I saw this bird on the 21st April 1883 on one of the open "plains" in the Fairmead Thicket; when it rose it flew a few yards only and re-alighted. I should not have known what it was if I had not been in the company of a well-known naturalist. A young bird of this species was captured on the borders of the Lower Forest.

Golden Plover. Not uncommon in winter. A flock of more than a hundred frequented Wanstead Park and the fields adjoining in company with lapwings and fieldfares in March 1883. Flocks also frequent Thornwood Common. Many of the male birds showed much black about the neck and breast.

Gray Plover. Has been found by the Roding in late autumn. It is generally considered a strictly maritime bird, and therefore its occurrence in the Forest is remarkable. One procured by the late Mr. Doubleday of Epping is preserved in the British Museum.

Lapwing. Frequently seen in flocks about the fields near

Wanstead Park and elsewhere. In the "Sixties" they used to breed on Fairmead and near the Wake Arms, but have ceased to do so. They are now seldom seen within the Forest, but there are some fields just outside its limits to which they still resort in spring.

Landrail or Corncrake. A summer visitor nesting and remaining till the middle of September.

Water-rail. Not so rare as is sometimes supposed. Owing to their silence and habit of creeping along the sides of brooks, and rarely taking to the water, they are not often seen; but they are sometimes very noisy in the breeding season.

Heron. These birds come to the heronry in Wanstead Park

HERONS.

at the end of January or beginning of February, and at once begin nesting operations. The old nests seem to require little repair. The number of nests for the last twenty years was as follows:—1882, 40; 1883, 36; 1884, 41; 1885, 45; 1886, 48; 1887, 46; 1888, 49; 1889, 51; 1890, 53; 1891, 50; 1892, 56; 1893, 61; 1894, 68; 1895, 55; 1896, 53; 1897, 48; 1898, 50; 1899, 50; 1900, 48; 1901, 45. The young birds may be heard in the nests early in March. The birds do not seem to mind the people, but I think there is some danger of their being evicted by the rooks, which have greatly increased of recent years. They are continually fighting the herons in the air, and instances are not unknown of a heronry being driven out in this way. The greater number of these birds leave the heronry towards the middle of summer, but a few remain throughout the year.

Woodcock. These birds are not unfrequent visitors to the Forest in the winter months, but the absence of springs and the hardness of the soil are unfavourable to them, and they are never numerous where cattle have access. They do not, therefore, generally remain long, but instances have been known of their staying through the summer and breeding.

Snipe. Often common by the Roding in Winter, and not unfrequent by Forest ponds. I saw a pair on the bare ground near Monk Wood, March 1897.

Jacksnipe. "I put up one by the Roding near Red Bridge, and, hiding myself, watched him return to the same

WATERHEN.

spot after a long flight, a well-known peculiarity of the bird."—A. L. I saw one on Piercing Hill in 1885.

Bittern. A specimen was killed a few years ago by the stream adjoining Wanstead Park.

Kittiwake Gull. An occasional visitor in stormy weather.

Common Sandpiper. Occasionally seen in spring and autumn by Leytonstone Pond, the waters in Wanstead Park, and by the Roding. I saw one last year by Baldwins Hill Pond.

Green Sandpiper. Occasional visitor in spring and autumn. One was shot in Mr. Barclay's grounds, Knotts Green. Two others shot near Epping are preserved in the British Museum.

Curlew. Seen several times and killed at Copped Hall.

Curlew Sandpiper. A specimen procured by the late Mr. Doubleday of Epping was presented by him to the British Museum.

Little Stint. Has once at least been procured near Epping. The specimen referred to is in the British Museum.

Gray Phalarope. "I set up a fine example of this beautiful bird, shot by the Rev. L. Cockerell at a brook, North Weald."—J. L. E. Another specimen was killed some years since at Wanstead Park.

Coot. Breeds in Wanstead Park, where they used to be numerous, but there are now very few left. They breed, however, at Birch Hall.

Moorhen. Most abundant in Wanstead Park and most of the Forest ponds.

Spotted Crake. Occasionally met with in the soft low-lying

WILD DUCK.

parts of the Forest, such as would be attractive to snipe. One may be seen preserved at the house of the head-keeper, Broad Strood Lodge.

Great-crested Grebe. "I observed a female bird on the Wanstead basin for several days in the spring of 1883."—A. L.

Red-necked Grebe. "In February 1877 one of these birds remained for almost a week on the basin in Wanstead Park. I watched it repeatedly with a good telescope. It was in fine plumage."—A. L.

Little Grebe or Dabchick. A great traveller notwithstanding his short wings. Breeds in the ponds at Wanstead and elsewhere in the neighbourhood. Like all the grebes, it constructs a floating nest, which is hard to distinguish from a mass of dirty weed.

Red-throated Diver. "In January 1877 I watched one

of these rare visitors on the Wanstead Basin. My telescope showed the speckling on the back quite distinctly; it was therefore probably a young bird."—A. L.

Wild Duck. Breeds annually in several parts of the Forest. There is generally a fair number that frequent Connaught Water, Wanstead Park, and other Forest Ponds, and as it is against the rule of the Committee to shoot them, they are delightfully tame.

Widgeon. May be seen occasionally in winter.

Teal. Several frequented the Wanstead Park waters in the spring of 1883, and in the following spring I saw a single bird on Connaught Water. We hope that it will take to breeding there. In hard weather it frequents the Roding.

Scaup. A flock of seven or eight remained on the Wanstead basin for several days a few winters ago. They were very shy.

Tufted Duck. Early in 1884 a male and three females frequented the upper pond in Wanstead Park for two months in company with the coots, and became almost as tame as they are. The male bird left us towards the end of February, but the females were still there March 7. They are excellent divers.

Pochard. Has been observed at Wanstead. On two occasions a single bird visited my pond and remained some weeks, consorting with the tame ducks, and becoming himself very tame. A single bird came to the Wanstead ponds early in the season some years ago and remained more than a month. At first he was very wild, but towards the end of his stay became tamer. Later on probably the same bird returned and kept company with the tufted ducks, acquiring their tameness. He went away with the male tufted duck towards the end of February.

Goosander. A flock of fifteen in immature plumage remained on the Basin at Wanstead for five days during the winter of 1885. "It was an interesting sight to watch the graceful fishing operations of these birds with a good telescope. After swimming in a compact company for a considerable time, they would all suddenly commence diving, probably having come over a shoal of fish, for many would be seen emerging with a fish in their bills, and, if one was too large to be immediately swallowed, a scramble would take place, and it would change beaks several times before being finally disposed of."—A. L.

Wild Geese. Flocks seen flying overhead during the winter months, especially brent, and gray of some species.

EPPING FOREST

THE ENTOMOLOGIST IN THE FOREST

BY W. COLE

"O happy living things! no tongue
Their beauty may declare."

As a locality for insects, Epping Forest suffers from at least two disadvantages—proximity to a great city with its smoke and teeming population, and the nature of most of the soil. The cold London Clay comes to the surface over by far the greater part of the district, and although patches of gravel appear in places, there are but few of the warm, bosky, sandy spots which flowers and insects so greatly love. A factor also to be reckoned with, to the sorrow of the ornithologist as well as of the entomologist, is the almost complete effacement by game-preservers of the raptorial birds, and the consequent undue abundance of the insect-destroying groups. Whatever theories may be advanced as to the effect of the attacks of insectivorous birds in controlling or enhancing the plague of certain kinds of caterpillars, there can be little doubt that pupæ which are laid up for the winter near the surface of the ground, or on palings, branches and trunks of trees, are destroyed to an extent which can only be truly appreciated by watching a tit or a creeper systematically clearing such nestling-places of their insect lodgers. Yet in spite of these drawbacks, Epping Forest has for years provided a happy hunting-ground for generations of entomologists, and the results of their researches, extant in faunal lists in almost all departments, have created for our woodlands quite a reputation among field naturalists. When it is remembered that Monk Wood is but twelve miles, as the crow flies, from the Royal Exchange, we may rejoice that it still gives shelter to such notable insects (to mention the moths only) as *Macroglossa fuciformis* (the Bee-hawk Moth), *Demas coryli* (the Nut-tree Tussock), *Psilura monacha* (the Black Arches), *Stauropus fagi* (the Lobster Moth), and the noble Emperor Moth (*Saturnia carpini*), alike remarkable in the caterpillar and moth stages, and in the wonderful structure of its cocoon.

The outspread of the birch tree in parts of the Forest will tend to the advantage of the entomologist, as it is a favourite pabulum with many species of almost all orders. As a possible result of this growth of birch woods, the beautiful and rare Large Emerald Moth (*Geometra papilionaria*) and the pretty Orange Underwing Moth (*Brephos parthenias*), whose lively flight is so welcome in the spring woodlands, have been noticed during the last two years in a part of the Forest not nine miles from the centre of London.

In common (alas !) with the experience of naturalists in most counties of England, our Forest butterfly fauna seems doomed to diminish; yet forty species (two-thirds of the British list) have been observed within the last fifteen or twenty years, and quite thirty-four butterflies may be expected to gladden the eyes of the observant rambler in and around the forest in this year of grace 1897, provided we have, as we wish for and deserve, perfect Diamond Jubilee weather. A few butterflies, such as the two Silver-spotted Fritillaries (*Argynnis selene* and *euphrosyne*), the Large Tortoise-shell (*Vanessa polychloros*) and the Dingy Skipper (*Nisoniades tages*), seem to be increasing in numbers; while the great growth of the holly in many of the woods is undoubtedly inducing the charming little Azure-blue Butterfly (*Lycæna argiolus*, whose larvæ feed on the flowers of the evergreen bushes) to enliven the sun-lit glades on April and May mornings more commonly than of yore. The abundance in most years of the Painted Lady (*Vanessa cardui*), formerly a rare butterfly with us, is a phenomenon widely observed of late in the South of England, and is not one peculiar to our district, pleasing as it is.

The forest, like all old woodlands, shelters hundreds of species of moths, both large and small, but the habits of many of these are very retired, and the insects are rarely seen unless specially sought for by those acquainted with their haunts. To enumerate even a tithe of them would fill this chapter with a dry list of Latin names.[1] A few which can hardly fail to be seen by the observant rambler in and about the forest may be mentioned. On the open heathy ground north of the Loughton Camp, on warm June afternoons, the wild dashing flight of the Fox Moth (*Lasiocampa rubi*) often attracts attention, and its comfortably clothed caterpillar may be found on the ling stems, by the sides of the pathways, in the late summer and autumn. On the outskirts of the forest the males of another great moth, the Oak Egger (*Lasiocampa quercus*), belonging to the same genus as the Fox Moth, fly with almost equal swiftness in the sunshine; they are in search of the females, which are sluggishly inclined, and are quietly nestling amid the herbage. The mysterious attractive power of the females (known as "assembling"), inherent in many moths, is extraordinarily developed in this species, and if a freshly emerged female is taken in a box to a lane or woodside where the moth is common, dozens of the males will speedily fly in from all directions and dart wildly about the imprisoned beauty. It is an astonishing spectacle, once seen never to be forgotten. The very handsome larvæ of this moth and of the Drinker (*Odonestis potatoria*) can readily be found stretched on grass or bramble stems in the hedges, basking in the

[1] A list of some of the rarer species was given in the earlier editions of this book (see pp. 100-102 of the third edition).

warm beams of a May-day sun. The last-mentioned caterpillar merits its popular name from its habit of drinking up drops of dew, which it does with evident gusto, lifting its head to swallow each mouthful with an action ridiculously like that of a hen similarly engaged. A careful examination of the blackthorn bushes may reward the searcher with the sight of the giant caterpillar of the Lappet Moth (*Gastropacha quercifolia*), but a sharp eye is needed for its detection, so perfectly does it harmonise with the branch or stem to which it closely clings. It affords a remarkable instance of the value of "protective resemblance" to so large a creature.

The great group of the Noctuæ can receive but a passing notice—the species are legion, but their sober colours and usually nocturnal habits conceal them from the ken of the ordinary visitor. A few, such as the handsome Red Underwing (*Catocala nupta*), occasionally fly by day, and the latter is often mistaken for a butterfly, so brilliant is its appearance when the sunshine gleams upon its wings as the moth dashes across a wood opening. The Red Underwing is very common in Lodge Bushes, and other parts of the district. A Noctua, rare all over Europe (*Erastria venustula*), was at one time thought to be peculiar to Epping Forest. The first specimens were taken there about 1830; these remained unique until 1845, when Mr. Doubleday found this pretty moth abundantly, *one June day only*, near Epping. It was then again lost, to be rediscovered quite lately; it is now observed almost every season in Monk Wood, but scarcely anywhere else in the three kingdoms!

The Geometræ and Tortrices also comprise exceedingly numerous species of the most varied and interesting habits; the groups are of importance to the forester, because they include those species of moths which are the main cause of the wholesale destruction of the foliage of the trees in parts of the forest in the early summer which has become so noticeable of late years. The species which work the havoc are mainly five in number, viz. *Hybernia defoliaria*, *H. aurantiaria*, *H. progemmaria*, and *Cheimatobia brumata* among the Geometræ, and *Tortrix viridana*, one of the Tortrices. Several other species also assist in the ravages, but the quintet mentioned are the greatest enemies to the green-wood.

The life-histories of the four Geometers are similar. The moths emerge in the autumn or winter from the pupæ, which are subterranean. The female moths are nearly or quite wingless; they can therefore only get at the twigs of the trees whereon to deposit their eggs by crawling up the trunks. The males have ample wings, and are to be readily found at night by the aid of a lantern. From the eggs laid on the twigs emerge at the time of bud-opening countless hordes of "looper" or geometriform caterpillars eager to begin their attacks on the tender foliage of the forest trees.

The habits of the pretty *Tortrix viridana* are different. Although so destructive to the oaks, we cannot refrain from admiring it as quite a little woodland gem, bright apple-green, with pure white underwings. The eggs lie dormant during the winter and hatch out in the spring. The young caterpillars form habitations for themselves by twisting up the oak leaves into little cylinders, inside which they find security and food. When full-fed the pupa state is assumed within the tube, and the moth emerges in July.

This little *Tortrix* and *Cheimatobia brumata* are mainly responsible for the defoliation of the *oaks*. At the slightest movement of the branches the *Tortrix* slips out of its tube, and the Geometers also quit hold, both hanging down a few feet by a silken thread until the danger is past, or if dashed to the ground by a strong breeze they ultimately remount the tree trunks, never ceasing as they travel to lay down their silken threads. The trunks and branches of the oaks thus frequently get quite overlaid with an unsightly matting of silk, and the spectacle of the defoliated branches, from which countless caterpillars sway by silken threads, has of late years become a sight only too frequent and distressing in our spring woodlands.

When we remember that each female moth lays some hundreds of eggs, it is manifest that there must exist some efficient controlling agents, or our woodlands would succumb to the plagues of caterpillars in a few years. Birds doubtless devour prodigious numbers of insects, but birds alone would be utterly unable to cope with the myriads upon myriads of such caterpillars as those we have above mentioned. It is to the various kinds of small hymenopterous flies, belonging to the Families Ichneumonidæ and Chalcididæ, that we must principally attribute this control. The eggs of these parasites are deposited within the bodies of the caterpillars, and the resulting grubs live and thrive on the substance of their victims, ultimately destroying them. It is a curious matter for speculation whether over-protection, and the consequent abnormal abundance of insectivorous birds, may not actually lead to an increase in some years of certain kinds of caterpillar pests. It is commonly observed by collectors that when caterpillars are "stung" they become sluggish, are easily shaken off the branches, and, by reason of abandoning their special methods of concealment, they are easily detected. Healthy larvæ are much more difficult to find. It is quite conceivable that birds may devour these sluggish, readily-seen larvæ, together *with the contained parasites*, leaving the more concealed and healthy ones to continue the race, and furnish forth a sufficient number of moths in November and December to provide another army of caterpillars in the spring. At the same time the ratio of the parasites to the caterpillars would have been disturbed by the unconscious selection of the birds in the previous season. This selection may go on until the caterpillars become a plague, or until some other natural controlling action (*which must exist*, although in some years latent) springs into force and restores the equilibrium between the parasites and their "hosts." What is the precise nature of this hidden controlling action, whether sudden variations in temperature or moisture, or (probably) the attacks of bacterial or fungoid diseases, is at present unknown. The study of the problem of the cause of the sudden increase or decrease of groups of insects is scarcely advanced beyond the purely speculative stage.

We have lingered too long, perhaps, with the Lepidoptera, albeit there are other divisions (such as the Tineina and Pyralides) equally worthy of the attention of the Field Naturalist, and we must now turn to some of the other groups of insects which find in the forest a congenial home.

The parasitical tribes of the Hymenoptera have been alluded to, but that great order includes many families. The Bees (Apidæ) and the Wasps (Vespidæ) are not so numerous in species or individuals as one would wish, owing to the fewness of sandy tracts and banks; there are, however, some interesting species, including the Hornet (*Vespa crabro*), and at least one species of aerial nest-making wasp (*Vespa norvegica*), the beautiful pendulous home of which is not very rare in gooseberry and currant plantations. The rambler can hardly fail to notice the large Wood Ant (*Formica rufa*) in the late summer, crawling over the paths and up the trunks of trees, and presenting a formidable front of distended mandibles and elevated head when irritated or alarmed. A little following up the lines of foraging ants will lead one to the nest beneath some hawthorn bush, externally a large conical heap of leaves and twigs, but within designed and reared with all the wondrous skill of some of Nature's best architects. (N.B. *Do not* disturb the nest when you find it.)

One or two species of the larger Humble-bees in the forest are very liable to be infested by that extraordinary internal parasite, *Sphærularia bombi*, belonging to the Thread-worms or Nematoda. When we were studying the subject years ago, it was found that 60 or 70 per cent, at least, of the specimens of *Bombus terrestris* caught while flying round the rhododendron flowers at High Beach were so infected. *En passant*, the parasites of insects is a fascinating subject, and one that opens out into quite a new field of observation for the entomological microscopist.

Still lingering with the Hymenoptera, we may well suppose that the whole range of zoology hardly affords a more wonderful subject for study than the life-histories of some of our common Oak-gall Flies (Cynipidæ). Galls are common enough in the forest, and the appearance of an oak tree with its leaves covered with the handsome cherry-like galls of *Dryophanta scutellaris*, or later in the summer with "Oak-spangles" (*Neuroterus lenticularis*) and "Silky-button" galls (*N. numismatus*), is sure to attract the attention of even the careless visitor. The "Oak Apple" gall (*Teras terminalis*) is sometimes extremely abundant, giving the oaks in May or June the appearance of trees in full fruit, and near Loughton the remarkable woolly galls of *Andricus ramuli* are often to be seen, looking like tufts of cotton caught on the twigs.

Oak-galls result from the deposition of an egg or eggs by the parent fly in the tissues of the developing bud or leaf. As the young grub grows, possibly some irritating secretion acts upon the tissues of the plant exciting to abnormal growth and excessive cell-division; but how mysterious is it that each particular species or generation of fly should on the same tree produce its own particular gall, often differing widely from those of allied species. And a still more wonderful tale anent gall-flies is that told by entomologists under the title of "Alternation of Generations." This principle will be best explained by an instance of it. The "Oak-spangle" gall (known as *Neuroterus lenticularis*) produces, in April, flies which are all females; these lay eggs in the buds of oaks, and, as the result, in May are to be seen the common currant-like galls either on the pendulous catkins or on the lower surface of the leaves. The flies from these latter galls are of *two sexes*, and they and the galls are known as *Spathegaster baccarum*. The female lays eggs in the oak leaves by piercing the cuticle with her ovipositor, so that in July is reproduced the "Oak-spangle" gall from which we started, and the cycle is complete. This alternation of asexual with sexual generations, and of strikingly different gall-formation, has already been traced in twenty or thirty species—our common Oak-apple gall (flies, male and female) alternates with a hard root-gall called *Biorhiza aptera* (females only); and the pretty cherry-like gall of the oak-leaf (alluded to above) is the agamous form of a little gall which is found in April on the undeveloped buds of the oak (known as *Spathegaster taschenbergi* (two sexes). A few gall-flies are only known in the agamous stage; for instance the Marble-gall of the oak (*Cynips kollari*), so common in the forest. Tens of thousands of the flies have been bred from these galls, but all females; no one has yet recorded a male!

We hope that some of our readers will take up the study of galls; the observation of their life-histories will provide an admirable lesson in biology; it is intended shortly to place in the Forest Museum at Chingford a series illustrative of most of the forest forms, for the information of students.

But the Cynipids are not the only gall-makers; in fact all orders of insects contain species which produce galls, and many more are formed by gall-mites (Phytoptidæ, Order Arachnida) which are not insects in the zoological sense. One of these is very common on the sloe, thickening and distorting the margins of the leaves (*Phytoptus attenuatus*); another occurs on Field Maple in the hedgerows, producing numerous little red pimples on the leaves (*P. myriadenum*); and the curious "Witches' Broom," so common on the hornbeams and birches, is an abnormal growth initiated by the attacks on the buds of a species of *Phytoptus*. The Diptera (two-winged flies) furnish several gall-making genera, particularly in the family Cecidomyidæ or Gall-gnats—tiny creatures, and yet a destructive race, for to the family belong the dreaded wheat-midge, and many species which distort leaves and flowers. A remarkable form may be found commonly in Monk Wood, small hairy tubular galls on the upper surface of the beech leaves, greenish at first and afterwards brown (*Hormomyia piligera*). Another species affects the flowers of the Sneezewort (*Achillea ptarmica*), and is interesting because it was first observed in Essex in Epping Forest (*Hormomyia ptarmicæ*, figured in *Essex Naturalist*, i. p. 177). In some seasons it is common on Fairmead.

The Diptera are a vast race (Britain contains about 3000 species, of which probably one-half occur in and around our forest), and it would be manifestly impossible even to enumerate the tribes. Epping Forest is almost unexplored in this department; the only recent paper on the forest species being that by Mr. E. Brunetti in the *Essex Naturalist* for 1890. No one can overlook the Diptera; they occur everywhere. Among the families most likely to be noticed by the casual observer are the Tipulidæ ("Daddy Long-legs"), of which several handsome species occur; the larvæ feed on rotten wood or roots of grass, and are often extremely injurious to farmers in Essex; and the Syrphidæ, or "Hoverer-flies," excessively active, brightly coloured creatures, common in the woods and lanes; many of the larvæ are *Aphis* eaters. The great family of the Muscidæ can only be mentioned—to this belong many of the commonest of known insects, including the House-fly and the Flesh-flies, but the group is very obscure and difficult to study. The grubs of certain species of the group Tachininæ are internal parasites of the larvæ of moths, and greatly aid the Ichneumons in keeping in check some of the larger caterpillars. Some Diptera in their early stages are leaf-miners; one cannot walk in the forest without noticing how profusely the leaves of the hollies are blotched with the whitish-looking mines of a species of *Phytomyza;* the larvæ can readily be found inside the burrows.

The Neuroptera (net-winged insects) are well represented, and quite two-thirds of the British species of Dragon-flies (Odonata) occur in the forest and the adjacent water-meadows; it falls little short, indeed, as a locality for these noble insects, of New Forest itself. Every one must have noticed the common *Platetrum depressum* (condemned to totally unmerited obloquy by reason of its rustic *sobriquet* "Horse-Stinger") with its bright cerulean body (in the male), as it hawks for flies over the forest pools, or rests on some outlying twig with its glistening wings outspread, ready to dart away at the slightest movement. In the ponds may be found the ugly larvæ, ferocious, but very deliberate in their movements; and high and dry on the rushes are the exuviæ (or "skin-cases") of the pupæ, left by the Dragon-flies as they exchanged their long sluggish aquatic life for that of one of the freest denizens of the air:

> To-day I saw the dragon-fly
> Come from the wells where he did lie.
> An inner impulse rent the veil
> Of his old husk: from head to tail
> Came out clear plates of sapphire mail.
> He dried his wings: like gauze they grew;
> Thro' crofts and pastures wet with dew
> A living flash of light he flew.

Another notable species in the forest is *Libellula quadrimaculata*, a very beautiful insect which is celebrated on the Continent for its migratory propensities, hordes of many millions having been observed in full flight. Epping Forest is one of the few localities for the grand *Anax formosus*, the largest European Dragon-fly, measuring nearly three inches in length, and nearly four in expanse of the wings. And a commoner species, but little inferior in size, *Æschna grandis*, may frequently be met with in the narrow glades in late summer, flying with all the celerity and ease of a swift, and moving as readily backwards as forwards. What a pattern for the inventors of æroplanes! The lovely, but vicious, "Demoiselle" Dragon-flies, *Calopteryx virgo* and *C. splendens*, are common on the Lea and Roding. We can but mention the group of the Caddis-flies (Phryganeidæ) whose aquatic larvæ, each in its own specially constructed tubular dwelling, are common in pools and streams; the beautiful Lace-winged Flies (Hemerobiidæ) so commonly to be beaten out of the trees and bushes, where the larvæ have battened on the Aphides; and the active Scorpion Flies (Panorpidæ), formidable-looking but harmless; the two British species *Panorpa communis* and *germanica* are common in our woods and hedgerows.

Nor will space allow of more than a passing reference to the families of Hemiptera, abundant as they are in the forest. Among the true Hemiptera (Heteroptera) we find many examples of the brightly coloured and active Plant-bugs (Pyrrhocoridæ and allied families), the Water-bugs including the Water-boatmen (Notonectidæ) and the Water-scorpion (*Nepa cineraria*) frequent in sluggish streams. In the sub-order Homoptera are the active Frog-hoppers (Cercopidæ) connected in popular lore with the cuckoo; countless hordes of Plant-lice (Aphidæ); and the very curious little "Powder-flies," Aleyrodidæ, several species of which are common among the beeches, and which were formerly mistaken for moths.

Among the Orthoptera, the Crickets (Gryllidæ) and the Grasshoppers (Acridiidæ) must be left to chirp their own praises, and there are several species to join in the chorus. Every one knows the Earwig, but there is a small species, *Labia minor*, which is commonly seen flying in the hot sunshine along roads or near farmyards. At least one species of Cockroach (Blattidæ) is found in the forest, *Ectobia lapponica*, which often comes to "sugar" spread on trees to attract Noctuæ; and but few can have failed to notice the pretty little Locustid, *Meconema varium*, pale green in colour, and with long graceful antennæ, and very capable hind-legs, which is often seen on trunks, or is beaten from the oaks and hornbeams in the late summer by collectors of caterpillars.

We have left the Beetles (Coleoptera) until the last; it would be impossible in a few paragraphs to give an adequate idea of the varied forms of this vast group which occur in the Forest. Beetles are found in every conceivable situation; some live under leaves and rubbish, and are nocturnal in their habits, as, for instance, the Geodephaga or Ground-beetles, and many of the Brachelytra, or "Cock-tail" beetles, to which group the common "Devil's Coach-horse" (*Ocypus olens*) belongs. Many occur in the open, like the brilliant but ferocious Tiger Beetle (*Cicindela campestris*), which may be seen on hot days darting on the wing with amazing rapidity in search of insect prey in the heathy ground near High Beach and Theydon. The tubular burrows of the larvæ may be seen in sandy pathways and banks. In these the inmate, a veritable British Ant-lion, awaits near the entrance the casual visit of some small insect, which is at once seized in the formidable jaws, with never a chance of retreat allowed. Another interesting predacious beetle is the somewhat local *Calosoma inquisitor;* this devours the caterpillars on trees, and it used to be quite common on the pollarded hornbeams in the forest near Woodford.

There are many families of Water-beetles (Hydradephaga), Plant-beetles (Phytophaga), containing some of the most brilliantly coloured of the order, and Lamellicornia to which the Rose-beetle (*Cetonia aurata*) and the great Stag-beetle (*Lucanus cervus*) belong. Another Stag-beetle, much smaller, but with a longer name (*Dorcus parallelopipedus*), is not unfrequently observed. To this group belong also the Geotrupidæ or Dung-beetles (Shakespeare's "Shard-borne beetles"), of which several species frequent the Forest, and whose amazing engineering feats in sinking deep cylindrical shafts in the solid soil for the reception of the nidus of dung in which the egg is placed, always command admiration.

A pretty weevil, *Rhynchites betuleti*, may be selected out of the vast array of the Rhynchophoræ or "Snout beetles" because of its curious habit of partially shearing off portions of the birch leaves, then rolling the pieces up like cigarettes; these are receptacles for eggs, and being left attached to the cut leaves, they ultimately turn brown, and are conspicuous objects on the young birches in Lodge Bushes and elsewhere.

Flowers attract numerous beetles of varied families; give a tap on a bright day at the end of May or June to a blossoming branch of hawthorn, and you will wonder at the lively company of beetles that have fallen into the open umbrella placed to receive them: Click-beetles (Elateridæ), Weevils (Curculionidæ), Telephoridæ ("Soldier-and-Sailor" beetles), Lady-birds (Coccinellidæ), species of *Melachius* in brilliant green and red coats, and scores of other interesting forms can thus be observed. Among them, if you are lucky, will be one or two species of *Clytus* (Wasp-beetles), and other Longicorns, *Leptura*, *Strangalia*, and the little *Grammopteræ*, etc. These long-horned, timber-loving beetles *ought* to be numerous both in species and individuals in an ancient forest, and perhaps they will still increase to the delight of the entomologist, now that trunks of felled trees are to be utilised as seats and overthrown trees left *in situ*. *Aromia moschata*, the brilliant and odoriferous Musk-beetle, one or two species of *Callidium* and other handsome "Long-horns" occur, and the fine *Prionus coriarius* has been taken as a great prize at Buckhurst Hill.

But we must draw this long chapter to a close, hoping that the imperfect sketch given of the insect riches of our woodlands will at least attract some few readers to take up a delightful study. The entomologist truly should be the happiest of naturalists, if an almost infinitude of objects upon which to exercise his observing powers be a measure of his satisfaction. Britain probably contains more than 12,000 species of insects, and a large proportion are, or ought to be, found in and around Epping Forest. Each species has its own life-history, structure, colour and "habit," differing in these respects from its congeners; it possesses special means of concealment or attack, or of adaptation to its surroundings, and it bears a definite relation to other living things—all these are matters deserving and repaying patient observation and investigation. Years of devoted study by generations of naturalists have but lifted the fringe of the subject; there are wide fields open for further exploration, some as yet almost untrod. The most learned and observant entomologist recognises how little he knows, compared with that which is unknown and unsuspected, and truthfully he can but re-echo the words of the dying Newton—"I know not how I appear to others, but to myself I seem but as a child playing on the sea-shore, amused by finding now and again a smoother pebble or a brighter shell than ordinary, the while the great ocean of truth lies all undiscovered before me."

EPPING FOREST

POND LIFE OF THE FOREST

By Dr. M. C. Cooke

It seems hard to realise that there is a world of animal and vegetable life, which is scarcely discernible by the naked eye, and was quite unknown, except in a few particulars, to our ancestors of one hundred years ago. And yet the perfection of the microscope, and the accession of such a crowd of manipulators, has rendered the study of pond life, as well as that of other minute organisms, one of the features of the age. In works which treat of the natural history of any district it is no longer permissible to ignore the "infinitely little," and leave the noble army of field workers with the microscope out of consideration. It is undoubtedly true that there are a large number of earnest amateurs in science, who trouble little about our Forest quadrupeds or birds, except for their hairs or feathers, and the parasites which flourish amongst them; or for our trees and wild flowers, except for the spots and distortions of their leaves, and the objects which cause them.

The lover of nature is no less interested in the hidden mysteries of the ponds, pools, and swamps of the Forest, than in its beech and hornbeam, and is as anxious for the preservation of objects almost invisible, as of the monarchs of the Forest and the beauties of the glade. For such as these there are ample fields of research, without in the least detracting from the general utility of the Forest to those who care for none of these things. The water area of the Forest is considerable, in addition to the expanse of Connaught Water and the ponds of Wanstead Park. There are numerous shallow ponds scattered over the entire area, and there are besides countless pools of a yard or two in extent, which are just the localities in which the pond-hunter delights to revel. It is to the natural inhabitants of these "still waters" that we venture to dedicate an additional chapter. When these several ponds come to be generally recognised by distinctive names, the localisation of the aquatic fauna and flora will become more definite and complete.

Unicellular Algæ. The smaller algæ, mostly unicellular, or sometimes more complex, but not truly filamentous or threadlike, may be alluded to together, although we can only suggest a few of them. Foremost for size and beauty is the well-known *Volvox* which we have collected at Golding's Hill, and in other places, sometimes in company with the smaller *Gonium pectorale.* As the spheres revolve in the clear water they present one of the most beautiful objects under a low power of the microscope. There is no reason why *Eudorina elegans* should not be found in similar places, although we have no record of it nearer than Abridge.

The different species of *Pediastrum* are represented by two or three of the largest, which have occurred at Snaresbrook, and in pools on the Loughton side. Scarcely less attractive is *Apiocystis Brauniana*, hitherto only recorded for Snaresbrook, but possibly not uncommon elsewhere.

There are a great number of very minute species which cannot be seen at all in the field, even with a pocket lens, and which will have to be examined at home, after the day's collecting is done. To provide for this contingency some Sphagnum may be placed, with the water in which it was growing, in a wide-mouthed bottle, and probably it will furnish some hours with delightful employment. In this manner, species of *Scenedesmus* may be obtained, or of *Raphidium*, and possibly some of the more imposing forms of *Polyedrium.* Tufts of some of the filamentous Algæ, especially when rather old, may be treated in the same manner, and will often reward the collector with quite a harvest of minute species, which could not be secured by any other process.

Desmids. Forest localities for Desmids were given by A. H. Hassell some fifty-three years ago, and since that time the improvements in microscopes, and the great accession to the ranks of observers, have increased the interest which attaches to Forest localities. Of all Fresh-water Algæ the Desmids are the most attractive to the ordinary microscopist. Only a few of the largest can be distinguished by the naked eye, but the beauty of most of them can be discerned by the use of low powers. The most satisfactory places to search are swampy ground, amongst bog moss, such as some localities about High Beach, the slopes in front of the King's Oak, and numerous other places in the Forest. Most of the species of *Closterium* are large, and can be seen in the collecting-bottle held up to the light. *Closterium lunula* at High Beach; *Closterium Ehrenbergii* near Chingford; *Closterium striolatum* on the Loughton side; and other species at High Beach, Snaresbrook, and Walthamstow. The large rounded fronds of *Micrasterias rotata* and *Micrasterias denticulata* occur in boggy pools about High Beach, Wood Street, and elsewhere. *Euastrum oblongum* and *Euastrum crassum* are about the finest of the genus, and these have been obtained at High Beach. Of the species of *Cosmarium*, such as *Cosmarium margaritiferum*, *Cosmarium botrytis*, are the most conspicuous, and these have been obtained at High Beach and Wood Street. One or two species of *Xanthidium* have been collected at High Beach.

Smaller species, such as those of *Staurastrum*, *Arthrodesmus*, and the more minute species of *Cosmarium*, are by no means uncommon. These will sometimes be found in conjugation. More than fifty different species may be relied upon, large and small, in the Forest district.

Filamentous Algæ. It is not uncommon to call all the forms of filamentous Algæ by the general name of confervæ, but this is an error which should be avoided. Most of the Algæ which have a threadlike form consist of a great number of cells joined end to end, and are at first attached to some fixed object, from which they are at length severed, and float freely in the water. They are of many kinds, and some of them of great interest, especially when they exhibit their fructification. Foremost in interest are the species of *Spirogyra*, in which the green endochrome is arranged in each cell, in the form of one or more spiral bands. A dozen or more species may be met with in the pools and clear ponds scattered over the Forest. One of the largest is *Spirogyra nitida*, and about as common as any of them; when in fruit the zygospores are very large. *Spirogyra longata* occurs at High Beach, as well as at Snaresbrook. *Spirogyra porticalis* is by no means uncommon, as also *Spirogyra flavescens* at High Beach and Wanstead Flats. Species with thinner and more delicate threads, such as *Spirogyra Weberi*, at Walthamstow and High Beach, and *Spirogyra tenuissima* at the same places. It is to be regretted that for some of the rarer species, the special ponds in which they have been found have not been indicated.

In *Zygnema* the endochrome is not arranged in spiral bands, but forms two star-shaped or globose bodies in each cell. Three species have been found at High Beach, and these are *Zygnema cruciatum*, *Zygnema Vaucherii*, and *Zygnema anomalum*, the former also at

Snaresbrook. They are not so common in the Forest district as are the species of *Spirogyra.*

The species of *Mesocarpus*, when in a barren state, have slender threads which resemble those of *Spirogyra* in a depraved condition, with the bands of chlorophyll zigzag, but not spiral. They are not at all common in this locality, but *Mesocarpus parvulus* has been found in the Forest, whilst *Mesocarpus pleurocarpus* occurs at Walthamstow. *Staurospermum* has threads equally slender, but are most interesting when they have conjugated, which they do at right angles so that the spores are four-cornered. *Staurospermum capucinum*, as well as *Staurospermum gracillimum*, are found in pools and ponds at High Beach, floating on the surface, and generally mixed with other Algæ.

Simpler forms of the threadlike Algæ, which are the true Confervæ, are often found floating on the ponds, sometimes in large masses; especially in the case of *Cladophora flavescens* and *Cladophora fracta*, which are common in most of the pools at High Beach, and elsewhere.

Of more interest are the numerous species of *Œdogonium*, the life-history of which is strangely attractive. The barren threads do not appear to differ much from a common conferva, except that the cells are longer and some of them present transverse parallel lines across the cells near one extremity. The cause of this is found in the manner of growth and development. They should be examined when in fructification, and then the floating tufts no longer appear green, but brownish, and uninviting. *Œdogonium vernale* is found at High Beach, *Œdogonium crassiusculum* on the Loughton side, *Œdogonium capillare*, *Œdogonium Boscii*, and *Œdogonium pluviale* at Walthamstow. That singular ally, *Bulbochæte setigera*, has been collected at Snaresbrook, and in some of the more northern ponds.

Two very elegant Algæ are often to be found scattered amongst other species, but never in great quantity, which are great favourites as microscopical objects. These are *Draparnaldia glomerata*, and *Draparnaldia plumosa*. Not only have we found them at Snaresbrook, but at Golding's Hill, and in other little ponds, especially such as contain *Volvox globator*.

Generally speaking, the Forest is not a good locality for the species of *Oscillaria*, but two species, *Oscillaria tenuis* and *Oscillaria tenerrima*, will be met with in ditches on the Chingford side. They are not so particular about clean water.

Protozoa. Within the past few years upwards of a hundred species of the Protozoa have been collected and identified from the pools of the Forest. Amongst these are some of the simplest forms of animal life, such as the *Amœba* and species of *Actinophrys*, and some six or seven species of *Difflugia*. More highly developed forms, long known under the general name of Infusoria, include many objects well known to the microscopist, which are not absent from the Forest. These include the Bell Animalcules or Vorticellæ, attached by long retractile threads to shells and other objects, constantly jerking up and down. Species with branched peduncles, known as *Epistylis*, are represented by some four or five species. Then again there are the Stentors, which are the largest of the Infusoria, of a funnel shape, with a fringe of delicate hairs around the mouth, for ever in motion. There is *Stentor polymorphus* in the southern ponds, with *Stentor mulleri*, and sometimes *Stentor niger*. To these may be added, as of universal interest, such things as *Vaginicola crystallina*, and the free-swimming *Stylonichia lanceolata*, and *pustulata*, as well as *Paramecium aurelia*, *Trachelius ovum*, and *Bursaria truncatella*. Different forms of *Euglena* are always to be found in immense numbers, so that not only *Euglena viridis*, but also *Euglena acus* and *Euglena longicauda*, as well as *Euglena deses*, may be obtained from nearly all parts of the Forest.

Hydrozoa. Of all the animals which inhabit ponds and ditches there are none to surpass the Fresh-water Hydra in interest, not only as representing the oceanic Hydrozoa, but from its remarkable history and possibilities. The green Hydra is the most common, and has been obtained in the southern part of the Forest, attached to aquatic plants. It has the advantage of being so large that it may be watched with the naked eye, expanding and withdrawing its arms, or tentacles, and catching its prey. There are also a brown and an orange-brown species, sometimes found mixed with the green species. No reference need be made here to the remarkable powers possessed by this little animal—of its being able to turn itself inside out, to produce little hydras by buds from its own body, and to suffer splitting into two halves, so that each half will heal up again and become a separate individual. All of this is now a matter of history, and has made the Hydra a classical object to the naturalist.

Polyzoa. No objects are more earnestly sought by the pond-hunter than the Fresh-water Polyzoa, of which two species at least occur in the Forest district. These are, *Fredericella sultana*, obtained near Chingford and Snaresbrook; and the other is *Lophopus crystallinus* from Chingford. Undoubtedly they are somewhat rare, and must be carefully hunted for, but, when found, will amply repay all the trouble. They attach themselves to aquatic plants, such as duckweed, in such manner as to be turned away from the light. When at rest, these Polyzoons expand their tentacles and present a most attractive appearance, but when these are withdrawn, they only resemble a lump of jelly.

Rotifera. The Rotifers, or wheel-bearers, are, as a whole, the most attractive of pond animals, inasmuch as some of them may always be found, and they present a very great variety of forms. Some of them are fixed, but the greater portion are free swimmers. Possibly there are more than three hundred known British species; at any rate nearly one hundred and fifty have been found and recorded in the still pools of the Forest. Those in the southern portion have been more explored than in the north, and from Hale End, Wood Street, Snaresbrook, and Chingford, most of the exploring expeditions have started. Of the fixed species none are better known than *Melicerta*, of which the commonest, *Melicerta ringens*, has been found about Snaresbrook, Wood Street, and Chingford. *Melicerta conifera* is recorded for Snaresbrook and Wood Street. Still more beautiful are the species of *Floscularia*, of which no less than five have been found—*Floscularia ornata* and *Floscularia cornuta* in most of the southern ponds, *Floscularia coronetta* at Snaresbrook and Chingford, *Floscularia edentata* at Snaresbrook, and *Floscularia campanulata* at Snaresbrook, Chingford, and Loughton. *Stephanoceros Eichhornii* has also been found in nearly all the southern waters.

Of the free-swimming species a mere list would be a formidable one, with nine species of *Anurea;* seven species of *Brachionus;* ten species of *Euchlanis;* not less than five of *Furcularia;* five of *Mastigocerca;* five of *Metopidia* and *Notommata;* four of *Œcistes* and *Philodina*, with six of *Rotifer*, and five of *Salpina*. In such a rich fauna it would be difficult to indicate individual species as of special importance, although it would appear that there are two or three species which have not been found outside of the Forest area, such as *Calladina magna-calcarata* and *Pterodina cœca*. It is of interest that the floating colonies of *Conochilus volvox*, at one time common on Hampstead Heath, and now almost unknown there, have been collected from Chingford to Loughton.

Entomostraca. The last group of microscopical animals to which reference can be made are the *Entomostraca*, or, as vaguely termed, the water-fleas. In this group the Forest waters are evidently strong, and especially the waters of Wanstead Park, which latter have been almost exhaustively explored.[1] Nearly seventy species constitute the list for the Forest district, of which sixty have been found in the ponds of Wanstead Park, and about forty in other localities. The old *Daphnia pulex* is associated with four other species of *Daphnia*, four of *Ceriodaphnia*, and two of *Daphnella*. The original *Sida crystallina*, and *Bosmina longirostris* are still to the fore, with five species of *Alona* and two of *Alonella*. There are also *Diaptomus castor* and *Diaptomus gracilis*, with *Canthocamptus minutus*, and no less than twelve species of *Cyclops*. If we add to these four species of *Caudona* and six of *Cypria*, with about twenty species distributed over seventeen genera, we shall have a brief summary of the Entomostraca of the Forest area.

[1] See the *Journal* of the Quekett Microscopical Club for July 1893.

EPPING FOREST

PREHISTORIC MAN AND THE ANIMALS WHICH HE HUNTED

BY DR. H. WOODWARD, F.R.S., V.P.G.S.

THE history of Epping Forest is given in the first twenty-one pages of this little book. But to every such area there is a far longer *pre-history*, hidden away in the ground beneath us, which, like the records of ancient Nineveh, may at any time be brought to light by the excavator.

Two such fragments of prehistoric records of the Forest may here be given.

1. At the extreme south-east corner of the Forest, in the district of Forest Gate, Ilford, and Wanstead, more than two hundred years ago, and I believe down even to the present day, extensive areas have been excavated to a depth of 10 to 20 feet to extract the *brick-earth*, which forms the *sub-soil*, and with which the red bricks of the houses of Queen Anne's time and earlier, in various parts of London, were mostly built. I examined the district many times from 1860 to 1865 with Sir Antonio Brady, F.G.S., long time a resident at Maryland Point, Stratford, Essex. His fine collection of animal-remains from Ilford is now preserved in the British Museum (Natural History), Cromwell Road, S.W.

This brick-clay is accompanied by beds of gravel and alluvium, and rests upon the older Eocene and Chalk, containing marine organisms, whilst the *brick-earth* is a superficial *terrestrial* deposit, washed down by rains and floods from the surface of the higher lands, and reassorted by the river Thames, which has again left it behind in the broader and more peaceful portions of its course and that of its affluents, the Lea and the Roding.

Buried with this fine brick-earth are many land and freshwater shells and abundant remains of land-animals, the ancient inhabitants of the country at this early period.

These comprise, the British lion, the hyæna, the bear, the fox, the "mammoth" (*Elephas primigenius*), the last-named elephant most abundant and of all ages; the straight-tusked elephant (*Elephas antiquus*); three species of rhinoceros, all, like the elephants, belonging to extinct species, viz. *R. leptorhinus*, *R. megarhinus*, and *R. antiquitatis.* An extinct horse, *Equus fossilis;* the gigantic Irish deer (*Cervus giganteus*); the red-deer or a variety of *Cervus elaphus;* the reindeer (*Rangifer tarandus*); *Bison priscus;* the fossil ox, *Bos primigenius* (very abundant); the musk ox (*Ovibos moschatus*); the Saiga antelope[1] (*Saiga tartarica*); *Hippopotamus major;* the pouched marmot (*Spermophilus citillus*); the lemming (*Lemmus*); the Alpine or tailless hare (*Lagomys spelæus*), have also been met with. Besides the shells of *Helices*, of *Anodon*, and *Unio* are countless valves of *Cyrena fluminalis*, a shell no longer found in this country, but still living in the Nile, etc.

There can be no doubt whatever that all these animals were denizens, not only of the Thames valley, but of the greater part of England in early prehistoric times; nor can it be doubted that prehistoric man, who fashioned those rude flint implements found in the Thames valley gravels in so many different localities, also inhabited this country and was contemporary with this ancient fauna, most of which are now extinct, and the survivors are only to be met with in far distant Arctic or continental lands. Even the geography of the area has vastly changed, for at the time

[1] Found higher up the valley at Twickenham in a similar deposit.

when these animals lived here Britain was united to Ireland, and also to the Continent, and was not isolated as she is to-day.

2. I had the opportunity in 1868-69 to carefully examine the excavations for reservoirs then in course of construction by the East London Waterworks Company, near Lea Bridge, Walthamstow Marshes (see map B).

In this area the deposits excavated consisted of surface-soil, loamy clay, peat, shell-marl, coarse and fine sands, rounded and subangular gravels.

Land and freshwater shells belonging to some thirteen genera and twenty-six species were obtained from the shell-marl, whilst the peat and shell-marl also yielded abundant remains of animals, comprising great numbers of wolves and foxes; the beaver (*Castor europæus*); the wild-boar of all ages; red-deer, roebuck, fallow-deer, goat; great abundance of heads and limbs, etc., of small British cattle (*Bos longifrons*) [and some *rare* remains of the elk, the reindeer, and the great ox (*Bos primigenius*), perhaps belonging to the older deposit already referred to], and bones of an eagle.

If we except the great ox, the elk, and the reindeer, we are here dealing with a deposit far younger than is indicated by No. 1. All the animals belong to the existing fauna, and the only ones not living in this country now are the beaver, the wolf, and the wild-boar; these are, however, still met with in Europe (although the beaver is becoming rapidly extinct). The small breed of British wild cattle (*Bos longifrons*) were probably the ancestors of our small Welsh cattle, and were "improved" out in Roman and later times by larger and better continental stock introduced from France and Normandy.

All these animals have survived in this country down to historic times, as our written records prove. I dug out of the reservoir bank a well-made flint scraper with my own hands from a bed of dark loamy clay 3 feet below the surface; whilst my late colleague, Sir Wollaston Franks, of the British Museum, obtained from the workmen bronze spear-heads, bronze arrow-heads, a bronze knife; an iron sword (late Celtic), part of the bronze sheath of a (late Celtic) dagger; a pierced axe-head of stag's horn; a bone knife; a stag's horn club; some Celtic earthen pots, and a Kimmeridge coal or jet armlet.

Such is the record which "the ground beneath us" yields to our investigations. We see portrayed the ancient forest extending over the whole country in an almost unbroken tract from the walls of British and Romano-British London, far to the north and over all East Anglia. The glades are tenanted by herds of deer and wild cattle; the woods and marshy ground are the home of the wild-boar; the secret places of the forest, the retreat of wolves; the rivers are tenanted by the beaver, whose dams resulted in the flooding and destruction of many a forest tract, by blocking up and diverting the natural water-courses of the Lea and the Roding. In evidence of this we see the peat beds full of timber, some of which was no doubt cut down by this ancient forester, who left his own remains also behind to "ear-mark" his work.

The primitive natives had little acquaintance with iron and bronze, and made their weapons of the chase of stone, bone, and horn. But when invaded by their warlike foes, the wild but well-armed Northmen and Vikings (who raided the coast and ascended the rivers in search of native villages which they might plunder and from which they might carry off the weaker inhabitants for slaves), these marauders occasionally left a sunk or stranded vessel behind, and some of their dead comrades together with their bronze or iron weapons to furnish us with a record of their unwelcome visits. So also do the marvellous "Deneholes" in the chalk at Grays—doubtless used by the natives in prehistoric times as places of retreat for their women and children, and for the storage of their grain and household gear in times of invasion or of local warfare.

At a still earlier period man was represented merely by small scattered tribes, or families of nomads, moving along the river banks subsisting on fish and forest animals, clothed in skins of wild beasts, his weapons rudely-fashioned stone celts, flint-pointed arrows, and others made of bone: a veritable troglodyte, without a settlement and owning no cultivated plot of land, nor any domestic animal save the dog, man's earliest friend and companion. He hunted the reindeer, made pitfalls for the mammoth, and lived indifferently on the wild horse, the reindeer, and wilder cattle. He was not insular then, but belonged to the same type as the aborigines of Western Europe, and, like the cave-men of France, incised pictures of the animals he saw and hunted upon the bones left from his repasts.

The great herds of wild horses, of bison, reindeer, and elephants (preyed upon by the lion, bear, hyæna, and wolf) moved to and fro, according to the season, across what is now the shallow bed of the North Sea and the Straits of Dover dry-shod, for it was then an old land-surface even up to the coast of Yorkshire and down to the Channel Islands.

The gigantic Irish deer, a great variety of the red-deer, and the Norway elk, the musk ox, and the Saiga antelope were also indigenes, with the lemming and the tailless hare. Hippopotami wallowed in the Thames and the Cam, and were common around our coast, and went north even as far as Leeds in Yorkshire. The greater beaver (the *Trogontherium*) had his home here also, and rhinoceroses were likewise quite common inhabitants.

We thus see that our forest has not only its recent and its ancient history, but its *pre*-history taking us back to periods so distant that we are compelled to abandon Time's hour-glass, which, like the compass at the magnetic north, has ceased to act as a reliable guide, and we are left stranded on the banks of quaternary rivers flowing into prehistoric seas where time records and nautical almanacs are alike unknown.

EPPING FOREST

ITS TREES

> "A shadie grove not farr away they spide . . .
> Whose loftie trees, yclad with sommer's pride,
> Did spred so broad that heaven's light did hide,
> Not perceable with power of any starr;
> And all within were paths and allies wide;
> With footing worne, and leading inward farr."
>
> *The Faerie Queene.*

EPPING FOREST has no timber to show equal, in dimensions or antiquity, to the rugged and venerable giants in many of the forests of England. The oaks of Sherwood and the New Forest, and the beeches of Burnham, far surpass those of Epping. Probably few of the latter date back more than 400 or 500 years. This deficiency is owing to the timber right, which was originally entirely under the control of the officers of the Crown who were charged with the preservation of "His Majesty's vert," having been subsequently vested in many hands, and to the abstraction for the benefit of these persons of most of the valuable trees. The Lord Warden of the Forest, for instance, and many others had perquisites in this way, and many orders are found in the Court Rolls, such as the following, for trees to be given to ecclesiastics and others for the repair of their buildings and for public purposes. An order dated 1722, for felling trees for repairs at Greenwoods Park, provided that "Noe tree fitt for ye service of his Majestie's navy be felled under cover hereof." But it was due, even to a greater degree, to the destructive custom of *pollarding*, or cutting back the branches to the crowns every fifteen years, at seven feet from the ground, for purposes of fuel. The few woods of larger growth, such as Monk Wood and High Beach, which we possess in the Forest area are exceptions to the rule; and it will be observed

HIGH BEACH.

that even many of these have been pollarded at some former time, as will be seen from the branches all breaking from the stem at the same height from the ground, although they were subsequently allowed unlimited freedom. We probably owe these groves to the foresight of the monks of Waltham, or possibly to their sporting instincts. At least, it is recorded that "they were permitted to assart their lands in Woodford and many other places; and enclose them with a ditch and low hedge, that they might take of their woods at their pleasure, to hunt the fox, hare, and cat in the forest; that their dogs should not be expeditated," etc. I ought to add, however, that, as regards Monk Wood, the lord of the manor of Loughton is of opinion that its unrestricted growth dates from much more recent times. If this is correct, we may hope that, within a few generations, the same larger growth will prevail over the whole of the wooded area, provided only that the trees are thinned in a timely and judicious manner. The mop-head growth substituted by the process of lopping, for the natural shape, is not only destructive of all variety and grandeur in the timber, but owing to the lodgment of moisture in the crown, and the consequent rotting of the heart of the timber, is fatal to the health and long life of the tree, and weakens its resistance to gales. All kinds of trees were subject to this periodical mutilation except the crab-apple trees, which abound in some sections, and which were specially excepted on account of the deer, who greedily devour the bright-coloured, but to our taste bitter, fruit, when it falls in the autumn. While, however, we cannot but regret the effects of this merciless treatment in the past, we can afford to look back leniently on this ancient right, unique in its way, on account of the important part which it bore in the preservation of the Forest (see p. 16). It is now abolished for ever, and those who exercised it were compensated under the orders of the arbitrator. Henceforth it is one of the most important duties of those who have the control of the Forest, to see that the restoring power of nature has free play as soon as possible.

In most of the Forest groves a fair number of "spear" or straight-grown trees, especially oaks, have been suffered to grow unmutilated, and though the finest have been removed by those who had the right to do so in bygone times, and perhaps by some who had not, there are many left, of varying ages, up to 150 or 200 years, which, with proper attention, will make fine timber trees for those who come after us.

The lowest branches of the beech and hornbeam are cropped close by the deer and commoners' cattle. This causes a dense hedge-like

HONEYLANE HILL.

growth late in the summer, which holds its withered leaves throughout the winter months, until they are pushed off by the young growth, and shows rich masses of brown in the dull season.

Much devastation has unfortunately been caused, especially in the neighbourhood of the Wake Arms and Theydon, by fires, accidentally, or I fear in some cases mischievously, kindled, many acres of charred stems and blackened ground showing a melancholy record. I would invite the co-operation of all visitors in averting this serious evil, and remind them that the careless dropping of a cigar-light, when the herbage is dry, may irreparably destroy many acres of copse, and that wrongdoers may be deterred by a word of caution or by information given to the keepers.

Considerable areas of Forest land, which were wrongfully enclosed, were cleared of trees and cultivated for several years before they were again thrown open. I have dealt with the subject of planting elsewhere.

Of the thousands of Londoners who refresh themselves by a visit to our groves, comparatively few do so at the two periods of the year when they are most beautiful—early May and late October. To the tree-lover I strongly recommend the former month. It is then, when the first burst of spring takes place, that the distinguishing characteristics, and especially the *colours*, of each tree can be best seen. A few weeks later the yellow of the young oak foliage, the gray-green of the birches, and the burnished light-reflecting quality of the beeches are merged in one uniform dark green, lovely in its varied shapes and play of light and deep shadow, but monotonous as to colour. The blackthorn in April, the hawthorn in May, a month later the crab, and sheets of the water-violet and water-ranunculus, supply the masses of white without which no group either of flowers or trees is complete. Again in the "fall," when our glades are almost solitudes, who can measure the glories of the beech groves when they put on their first autumn touches of brown and gold, contrasting with the dark green of the more persistent oak, and with the saffron yellow of the birch and maple? No leaf-colouring can be found to surpass these in brilliancy. And yet these beauties are mostly wasted on the desert air. Let me then invite the visitor to extend his migratory periods at both ends.

In the following list I have described only those trees which are native to the soil. There are many others, such as the horse chestnut, Spanish chestnut, poplar, elm, walnut, sycamore,

Scotch fir, willows of various sorts, etc., of which a few individual specimens may be found in the Forest; but as they are not indigenous, but have either been planted or have seeded themselves from cultivated ground, I take no particular notice of them here. I have also omitted the hazel, as, although we learn that Gilbert de Ecclesia of Chingford "was obliged, by the tenure of his lands, to find a man to gather *nuts* for the lord of the manor," he would find it impossible to fulfil his bargain at the present day.

The **Oak** (*Quercus Robur*). There are two varieties of the common oak in England—*sessiliflora* and *pedunculata*. It is the former which abounds in the Forest. It is distinguished from the latter by the acorns being borne close to the stalk instead of on footstalks, and by smaller foliage. The largest trunk on the Forest ground stands just within the enclosure by Fairmead Lodge. It has been pollarded up to comparatively recent times, so that the spread of branches is not great, but it has a grand rugged stem 22 ft. 7 in. in circumference at 3 feet from the ground. On the rising ground beyond Connaught Water there is a fine tree of wide spread, known as Grimston's Oak, so named after the late Hon. Robert Grimston, the well-known cricketer, who first called attention to it, and at whose suggestion a space was cleared around it; but the oak of grandest dimensions in the Forest is the old pollard that stands by Fairmead Lodge. There are many picturesque oak pollards of great age in Lord's Bushes, but not only have they been lopped until recently, but they have been overshadowed by younger and more vigorous trees, and their decay thereby much hastened. The oaks are much injured in some seasons by a plague of the small green caterpillar[1] (*Tortrix viridana*), which devours the young leaf in May and June. In some seasons every vestige of foliage is eaten from many of the oaks, and the caterpillars hang in festoons from the bare branches, which are surrounded later in the year by dense clouds of the small moths. Although a second foliage is put forth many trees are seriously injured. It is perhaps indirectly caused by severe winters, which by destroying immense numbers of birds promote an unnatural increase in the insects upon which they prey. The jays, by destroying great quantities of eggs, befriend the caterpillars, and are partly responsible for the mischief. The birds have, however, now recovered their majority. The caterpillars are in opposition, and the foliage just now is less injured.

The **Beech** (*Fagus sylvatica*) is certainly entitled to the second place, if not the first, among the Forest trees for

[1] For the life-history of those and other destructive pests, see Mr. Cole's Chapter, p. 83.

its beauty and hardiness. It is easily distinguished by its smooth massive stem and light-green satiny leaves. From its habit of rooting close to the surface, and from its dense foliage, as a rule, nothing will grow under its branches; but the bare surface is very beautiful when it is carpeted in spring with the brown sheaths which have protected the buds during winter, and again by the falling leaves in autumn. Less patient of moisture than the oak, few good specimens are found in such stiff soils as Fairmead thicket; but on the drier hills it is crowding its rival out, and makes fine timber. It is as noble a tree as the oak, but not so long lived. The leaves are far less liable to destruction by caterpillars than the oaks. The "mast" or nuts are shed in great profusion, and are a favourite food of the deer. The best beeches in the Forest are found along the green ride in Epping Thicks. Unlike the groves of High Beach and Monk Wood, these have never been pollarded; and, as they stand somewhat thinly, they have not been punished by competing trees, but have made fine lateral growth which weeps nearly to the ground.

The **Hornbeam** (*Carpinus Betulus*) is a tree of less imposing growth than either of the preceding, but it is much more widely spread than either of them. No soil, however wet or dry, comes amiss to it; and, undiscouraged by the most ruthless pollarding, it at once "starts" again. It was consequently a favourite tree with the "loppers," and there are scarcely any good specimens in the Forest which have not been periodically beheaded. Epping Thicks possesses two fine "spears" of this species, but I should find it hard to point out more than one or two others. This cruel treatment is now put an end to in Epping Forest, and we may expect them to respond with their graceful horizontal growth. It is said to have been originally encouraged on account of the deer, who browse on its leaves, which are among the earliest to appear. These resemble those of the beech, but are less polished in appearance, and have a saw-like edge. The branches have a curious habit, of which many quaint specimens may be found in the Forest, of reuniting several feet from their point of separation and again forming one stem. The winged seed vessels of the hornbeam are a very pretty feature when the tree is covered with them. This effect generally follows a hot summer and is not produced every year.

The **Birch** (*Betula alba*). If the oak is the king of trees, this is the queen. While the beech is remarkable for the massiveness of its foliage, the birch is distinguished by its grace and lightness. Its white stem and fine spray-like branches are a lovely feature even in the depth of winter. Formerly it was not common in the Forest, but, from its enormous production of seeds, which are carried far on the wind by the little wing attached to them, it is spreading itself rapidly, and I anticipate that, in a comparatively short period, it will to some extent supplant the other trees. Wherever a clearance has been made, either intentionally or by accidental fires, if the soil be dry, it appears to spring up spontaneously. The rough open ground by High Beach Church has been quite recently covered by a charming grove of birches, which have probably sown themselves from the neighbouring gardens; and, in the case of a wide roadway cut through Lord's Bushes about 1860 with the intention of selling it in building lots, the crime is already hidden under a similar screen. The same thing is taking place in the open plain to the north of the Theydon road. Those who are now middle-aged will live to see the bare plain between the "Wake Arms" and Monk Wood and other openings, similarly restored by nature. The birch has two attributes which are specially valuable on the Forest. One is the fact that, at a very early age, it acquires a roughness of bark at the base, which gives it partial immunity from the effect of casual fires. The other is that it is very little browsed by cattle.

The **Maple** (*Acer campestre*) is found sparingly in the moister parts of the Forest. It is a small tree, and is not conspicuous until the autumn, when it turns to a fine golden yellow.

The **Common Hawthorn or May** (*Cratægus Oxyacantha*).—This is, next to the oak, one of the longest lived of trees. Its rugged stems assume picturesque forms, and are highly characteristic of the Forest, which it enlivens by its early foliage, by its masses of white flowers, and by its autumn berries. Waggoners, passing through the Forest to London, deck their teams with great branches of May blossom, but there is enough and to spare for all. When not overshadowed by other trees it grows in a dense bush, not more than 12 feet high, with a flat rounded head.

The **Blackthorn** (*Prunus spinosa*) forms the most impenetrable undergrowth in the Forest. It flowers earlier than the May, and is a welcome forerunner of spring. It is the parent of our plums and bullaces, which have been developed from it by ages of cultivation in our gardens.

The **Common Crab** (*Pyrus Malus*) is the wild representative of the apple. It is widely spread through the Forest. The blossoms, with which the tree is smothered in May, are highly ornamental. The deer feed greedily on the apples, which they eat whole, and do not seem any the worse for. For this reason the tree was not allowed to be pollarded. The stems are rugged and picturesque.

The **Bird Cherry** (*Prunus Cerasus*).—I believe I have seen this tree growing in the Forest, and have no doubt that it used to do so, but I cannot now put my hand on it. In any case it is scarce. It is gay in spring, and the autumn leaves colour richly.

The **Common Sallow or Goat Willow** (*Salix caprea*) forms a small bush in rough places and the hollows left by old gravel-pits. It produces the "goslings" which children are fond of gathering at Easter.

The **Holly** (*Ilex Aquifolium*), the chief evergreen we possess, grows freely throughout the Forest wherever the soil is tolerably dry. It is not much observed in the summer ; but when other trees are bare, from the shelter and seclusion which it affords, and from the deep green of its glossy leaves, it is invaluable. For some reason which I cannot explain it does not often bear berries in the Forest. This is as well, as otherwise large quantities of it might disappear at Christmas time. The deer browse on it in winter. Consequently, since those animals have increased in number, the holly has made less progress. The finest holly undergrowth is found in Lord's Bushes, and the Walthamstow Forest, where there are no deer.

Butcher's Broom (*Ruscus aculeatus*) is the only other evergreen indigenous to the Forest. There is not much of it, and what there is, is not very ornamental.

EPPING FOREST

ITS FLOWERING PLANTS

> " Flowers worthy of Paradise, which not nice art,
> In beds and curious knots, but Nature boon,
> Poured forth profuse on hill, and dale, and plain ;
> Both where the morning sun first warmly smote
> The open field, and where the unpierced shade
> Imbrowned the noon-tide bowers."—*Paradise Lost.*

THE flowers of the Forest, though for the most part common enough, are not on that account less attractive in the eyes of Londoners. Of the many thousands, who visit us on every sunny day in the summer, few return without securing some floral trophy, which it has given them infinite pleasure to gather.

Owing to its variety of soil and aspect, and the contrast of extremely dry positions with damp hollows, the Forest is, for its extent, particularly rich in its flora. Many of the flowers which reward the diligent seeker are rare or structurally remarkable. Among these may be mentioned the sundew, which catches small insects with its arrangement of gummy hairs, and nourishes itself on their dead bodies; the grass of parnassus, a beautiful white flower to be found in boggy places; the bog-bean; and the lily of the valley. The last named is uncommon, and if I mentioned its locality it might be exterminated; but as it does not flower here I trust it may be spared.

The plants which may be described as most characteristic of the Forest are the common brake fern, which is universal; the common polypody fern, which grows in the crowns of the pollard trees; the gorse; the petty whin; and the ling or heather, which cover the drier parts; the orchis, of which more than one variety used to be found beyond High Beach; and the ground ivy on the clay soil. The aquatic vegetation is a distinct feature, rich in deep greens and graceful forms. Of the water plants the common water-dock is one of the handsomest; the two kinds of bulrush, one tall and very slender, the other bolder in character; the water-lilies; the carices or water-grasses; the sweet-scented rush, with which in ancient times the Monastery floor was strewn; the water-violet; the ranunculus; and the mare's tail, are all full of beauty and character. But the real show is afforded by the masses of the commonest and most familiar flowers, each in their season, and especially of the flowering bushes. Thus in March we have the primrose, though unfortunately the stock of these has been sadly diminished by being dug up in bushels to be sold in London; in April the blue bells and white wood anemone, which may be gathered in quantity in the Walthamstow Forest, at the Cuckoo Pits, between Chingford and High Beach, and elsewhere; in May the hawthorn, crab apple, and gorse; in June the wild rose, water forget-me-not, and many others; in July the honeysuckle and tall flowering grasses, and in the pools the graceful bulrushes. In September the floral attractions yield to the more succulent charms of the blackberry, and until December the gorgeous autumn colours of the trees furnish many a lovely nosegay.

I am indebted to two excellent authorities for the following complete list,—to Professor Boulger, President of the Field Club for 1884, who has devoted much time and attention to the subject; and to Messrs. Ware of Tottenham, whose skilled florists are constantly perambulating the Forest, and who have taken much pains to furnish me with the complete results of their researches.

The less common, or more interesting, are marked with an asterisk.

***Clematis Vitalba.** Traveller's Joy. Epping. Dry places.
Anemone nemorosa. Wood Anemone.
Ranunculus floribundus. Water Crowfoot.
R. trichophyllus. Water Fennel.
R. peltatus, **R. hederaceus** } Water Crowfoots.
R. sceleratus.
R. Flammula. Lesser Spearwort. Marshes.
R. Ficaria. Lesser Celandine.
R. auricomus. Goldilocks.
R. acris, **R. repens**, **R. bulbosus**, **R. hirsutus** } Buttercups.
Nasturtium officinale. Water-cress.
Cardamine pratensis, **C. hirsuta** } Ladies' Smock. Moist places.
Sisymbrium officinale. Hedge Mustard.
Alliaria officinalis. Garlic Mustard.
Sinapis arvensis. Charlock.
S. alba. White Mustard.
***Alyssum maritimum.**
Draba verna. Whitlow-grass.
***Teesdalia nudicaulis.** Shepherd's-cress.
Capsella Bursa-pastoris. Shepherd's-purse.
Viola odorata. Sweet Violet.
V. sylvatica. Wood Violet.
V. arvensis. Wild Heartsease.
***Drosera rotundifolia.** Sundew. Marshes.
Parnassia palustris. Grass of Parnassus. Marshes.
Polygala vulgaris. Milkwort.
Silene inflata. Bladder Campion. Dry places.
Lychnis vespertina. Evening Campion. Sunny places.
L. diurna. Red Campion. Coppice.
Sagina procumbens, **S. apetala** } Pearlwort.
Arenaria trinervia.
A. serpyllifolia.
Stellaria Holostea. Greater Stitchwort.
S. graminea. Lesser Stitchwort.
S. uliginosa. Bog Stitchwort.
Mœnchia erecta.
Cerastium aquaticum. Water Chickweed.
C. glomeratum. Mouse-ear.
C. triviale.
C. semidecandrum.
C. arvense.
Malva sylvestris. Mallow. Dry places.
M. rotundifolia. Dwarf Mallow. Dry places.
***Hypericum androsæmum.** Tutsan. Thickets.
H. tetrapterum. Square-stalked St. John's-wort. Damp places.
H. perforatum. Perforate St. John's-wort.

Hypericum humifusum. Trailing St. John's-wort.
H. hirsutum. Hairy St. John's-wort.
H. pulchrum. Upright St. John's-wort.
***H. elodes.** Marsh St. John's-wort.
Acer campestre. Maple.
Geranium dissectum
G. rotundifolium } Wild Cranesbills.
G. molle
G. robertianum. Herb Robert.
Erodium cicutarium. Stork's-bill.
Radiola millegrana. Flax-seed.
***Oxalis Acetosella.** Wood-sorrel. Shady places.
***Euonymus europæus.** Spindle-wood.
***Rhamnus catharticus.** Buckthorn.

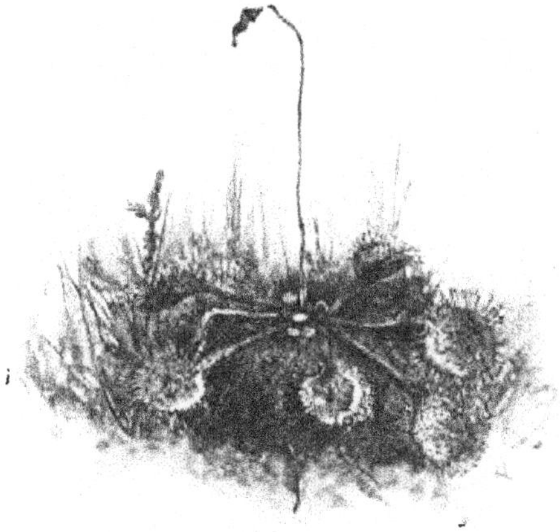

SUNDEW.

***R. Frangula.** Alder Buckthorn.
Ulex europæus. Furze.
U. nanus
U. Gallii } Dwarf Furze.
***Genista anglica.** Petty Whin. Heathy places.
***Cytisus scoparius.** Broom. Sandy places.
Ononis campestris. Rest-harrow.
Medicago lupulina. Nonsuch.
M. maculata. Spotted Medick.
Trifolium pratense. Purple Clover.
T. maritimum. Sea Trefoil.
T. repens. Dutch Clover.
***T. ornithopodiodes.**
***T. fragiferum.** Strawberry-headed Trefoil.
T. minus.
Lotus corniculatus. Bird's-foot Trefoil.
L. major. Greater Bird's-foot Trefoil.
Vicia hirsuta. Hairy Tare.
V. tetrasperma. Smooth Tare.
V. Cracca. Tufted Vetch.

G

Vicia sepium. Bush Vetch.
***V. Bobartii.** Small Crimson Vetch.
Lathyrus pratensis. Meadow Vetchling.
Orobus tuberosus. Bitter Vetch.
Ornithopus perpusillus. Bird's-foot.
Prunus spinosa. Sloe.
Spiræa Ulmaria. Meadow-sweet. Wet places.
Agrimonia Eupatoria. Agrimony. Sunny places.
Alchemilla arvensis. Parsley Piert.
Potentilla anserina. Silver-weed.
P. **reptans.** Creeping Cinquefoil.
*P. **Tormentilla.** Tormentil.
P. **procumbens.** Trailing Tormentil.
P. **Fragariastrum.** Barren Strawberry.
Fragaria vesca. Wild Strawberry.
Rubus Idæus. Raspberry.
R. Lindleianus
R. affinis
R. discolor
R. leucostachys
R. carpinifolius } Brambles.
R. pallidus
R. Koehleri
R. glandulosus
R. corylifolius
R. Cæsius. Dewberry.
Geum urbanum. Avens.
Rosa tomentosa. Downy Rose.
R. micrantha. Lesser Sweet-briar.
R. rubiginosa. Sweet-briar.
***R. canina.** Dog Rose.
R. arvensis. White Rose.
Cratægus Oxyacantha. Hawthorn.
***Pyrus malus.** Crab Apple.
var. **acerba.**
var. **mitis.**
*P. **torminalis.** Wild Service.
Peplis Portula. Water Purslane.
***Epilobium angustifolium.** Rosebay } Moist places.
E. hirsutum. Codlins-and-cream }
E. parviflorum. Small-flowered Willow-herb }
E. montanum. Smooth-leaved Willow-herb } Moist
E. tetragonum. Square-stalked Willow-herb } places.
E. palustre. Narrow-leaved Willow-herb }
***Circæa lutetiana.** Enchanter's Nightshade. Shady places.
Myriophyllum alterniflorum. Water Milfoil.
Bryonia dioica. White Bryony.
Montia fontana. Water Blinks.
M. rivularis.
Arenaria rubra. Red Sandwort.
Saxifraga tridactylites. Dry situations.
S. granulata. Meadow Saxifrage.
Chrysosplenium alternifolium } Golden Saxifrage.
C. oppositifolium } Marshes.
***Hydrocotyle vulgaris.** White-rot. Marshes.
Sanicula europæa. Wood Sanicle.

Sison Amomum. Hedge Honewort.
Bunium flexuosum. Earth-nut.
Æthusa Cynapium. Fool's Parsley.
Silaus pratensis. Pepper Saxifrage.
Angelica sylvestris. Wild Angelica.
Peucedanum palustre. Milk Parsley.
Heracleum Sphondylium. Cow Parsnip.
Daucus Carota. Carrot.
Caucalis Anthriscus. Upright Hedge Parsley.
C. nodosa. Knotted Hedge Parsley.
Scandix Pecten-veneris. Shepherd's Needle.
Chærophyllum sylvestre. Wild Chervil.
C. temulum. Rough Cow Parsley.
Conium maculatum. Hemlock.

DEWBERRY.

***Adoxa Moschatellina.** Moschatel. Shady places.
Hedera Helix. Ivy.
Cornus sanguinea. Dog-wood.
***Viscum album.** Mistletoe.
***Sambucus Ebulus.** Danewort.
S. nigra. Elder.
Viburnum Lantana. Wayfaring-tree.
***Lonicera Periclymenum.** Honeysuckle.
Sherardia arvensis. Field-madder. Open fields, fallows, etc.
***Asperula odorata.** Woodruff.
Galium Aparine. Goose-grass. Hedgerows.
G. Mollugo. Hedge Bed-straw.
G. verum. Yellow Bed-straw. Heaths.
G. saxatile. Heath Bed-straw.
G. uliginosum. Marsh Bed-straw.
G. palustre. Water Bed-straw.
Valeriana officinalis. Great Valerian. Shady places.
Dipsacus sylvestris. Teasel. On clay.
Knautia arvensis. Field Scabious.
***Scabiosa succisa.** Devil's-bit Scabious. Heaths.
Tussilago Farfara. Coltsfoot.
Bellis perennis. Daisy.
***Solidago Virgaurea.** Golden-rod.

Pulicaria dysenterica. Fleabane.
Bidens tripartita. Bur-marigold } Damp places.
B. cernua. Nodding Bur-marigold }
Anthemis arvensis. Corn Chamomile.
A. Cotula. May-weed.
A. nobilis. Common Chamomile.
Achillea Ptarmica. Sneezewort.
A. Millefolium. Milfoil.
Chrysanthemum Leucanthemum. Moon-daisy.
Matricaria Parthenium. Feverfew.
M. Chamomilla. Wild Chamomile.
Artemisia vulgaris. Mugwort.

TEASEL.

Filago germanica. Cudweed.
F. minima. Least Cudweed.
Gnaphalium uliginosum. Marsh Cudweed.
G. sylvaticum. Wild Cudweed.
Senecio vulgaris. Groundsel.
S. Jacobæa. Ragwort.
S. aquaticus. Waterwort.
Arctium majus }
A. intermedium } Burdocks.
A. minus }
Serratula tinctoria. Saw-wort.
Centaurea nigra. Black Knapweed.

Centaurea nigrescens.
C. Scabiosa. Great Knapweed.
Onopordon Acanthium. Cotton Thistle.
Carduus crispus. Curl-leaved Thistle.
C. lanceolatus. Spear Thistle.
C. arvensis. Creeping Thistle.
***C. palustris.** Marsh Thistle.
C. pratensis. Meadow Thistle.
Lapsana communis. Nipplewort.
Hypochæris radicata. Cat's-ear.
Apargia hispida. Rough Hawkbit.
A. autumnalis. Autumn Hawkbit.
Tragopogon minus. Lesser Goat's-beard.
Helminthia echioides. Bristly Ox-tongue.
Lactuca muralis. Wall Lettuce.
Leontodon Taraxacum. Dandelion.
Sonchus oleraceus, **S. asper**, **S. arvensis** } Sow Thistles.
Hieracium Pilosella. Mouse-ear Hawkweed.
H. vulgatum. Wood Hawkweed.
H. boreale. Broad-leaved Hawkweed.
Campanula rotundifolia. Hare-bell. Heathy places.
***C. hederacea.** Ivy-leaved Bell-flower. Bogs.
***Calluna vulgaris.** Ling.
***Erica Tetralix.** Cross-leaved Heath.
***E. cinerea.** Fine-leaved Heath.
***Vaccinium Myrtillus.** Bilberry.
***Ilex aquifolium.** Holly.
Fraxinus excelsior. Ash.
Ligustrum vulgare. Privet.
Erythræa Centaurium. Centaury. Sunny places.
***Menyanthes trifoliata.** Buckbean. Bogs.
Convolvulus arvensis. Bindweed.
***Cuscuta epithymum.** Lesser Dodder.
Lithospermum arvense. Corn Gromwell.
Myosotis palustris. Forget-me-not. Damp places.
M. cæspitosa, **M. arvensis**, **M. versicolor** } Scorpion-grasses. Dry places.
Solanum Dulcamara. Bitter-sweet.
***Atropa Belladonna.** Deadly Nightshade.
***Hyoscyamus niger.** Henbane.
***Orobanche major.** Great Broom-rape.
Verbascum Thapsus. Great Mullein.
***Digitalis purpurea.** Foxglove.
Antirrhinum Orontium. Lesser Snapdragon.
Linaria vulgaris. Toadflax.
Scrophularia nodosa. Figwort.
S. aquatica. Water Figwort.
***Limosella aquatica.** Mudwort.
***Melampyrum pratense.** Cow-wheat.
Pedicularis palustris. Tall Red-rattle, **P. sylvatica.** Dwarf Red-rattle, **P. sylvatica flore albo.** White variety } Damp places.
Euphrasia officinalis. Eyebright.
Bartsia Odontites. Dry places.

Veronica Anagallis. Water Speedwell.
V. Beccabunga. Brooklime.
V. Chamædrys. Germander Speedwell.
V. officinalis. Common Speedwell.
V. serpyllifolia. Smooth Speedwell.
V. arvensis.
V. Buxbaumii.
V. polita.
V. hederifolia.
Mentha aquatica. Water Mint.
M. agrestis.
M. Pulegium. Penny-royal.
Lycopus europæus. Gipsy-wort. Damp places.
Calamintha Clinopodium. Wild Basil.
Melissa officinalis. Balm.
Prunella vulgaris. Self-heal.
Nepeta Glechoma. Ground-ivy.
Lamium purpureum. Red Dead-nettle.
L. album. White Dead-nettle.
L. Galeobdolon. Yellow Dead-nettle.
Galeopsis Tetrahit. Hemp-nettle.
***Stachys Betonica.** Wood Betony.
S. sylvatica. Hedge Woundwort.
Ballota fœtida. Black Horehound.
Marrubium vulgare. White Horehound.
***Teucrium Scorodonia.** Wood Sage.
Ajuga reptans. Bugle.
Verbena officinalis. Vervain.
***Utricularia vulgaris.** Bladderwort } In water.
***U. neglecta.** }
Primula vulgaris. Primrose.
P. veris. Cowslip.
***Hottonia palustris.** Water-violet.
Lysimachia Nummularia. Money-wort. Wet places.
L. vulgaris. Loosestrife.
L. nemorum. Yellow Pimpernel. Moist shady places.
Anagallis arvensis. Scarlet Pimpernel.
***A. tenella.** Bog Pimpernel.
Centunculus minimus. Bastard Pimpernel.
Plantago lanceolata. Ribwort Plantain.
P. media. Hoary Plantain.
P. major. Greater Plantain.
***Littorella lacustris.** Shore-weed.
Chenopodium album. White Goose-foot.
C. Bonus-Henricus. Good King Henry.
Atriplex angustifolia }
A. erecta } Oraches.
A. deltoidea }
A. hastata }
Rumex conglomeratus }
R. sanguineus } Docks.
R. obtusifolius }
R. crispus }
R. acetosa } Sorrels.
R. acetosella }
Polygonum lapathifolium.
P. Persicaria. Persicarias.

Polygonum Hydropiper
P. minus
P. maculatum } Persicarias.
P. **aviculare.** Knot-grass.
P. **Convolvulus.** Black Bindweed.
***Daphne Laureola.** Spurge Laurel.
Euphorbia Helioscopia. Sun Spurge.
E. amygdaloides. Wood Spurge.
E. Peplus. Petty Spurge.
Euphorbia exigua. Dwarf Spurge.
Mercurialis perennis. Dog's Mercury.
Ceratophyllum demersum. Hornwort.
Callitriche verna. Water Starwort.
Urtica urens. Small Nettle.
U. dioica. Common Nettle.
Ulmus campestris. Common elm.
Salix aquatica. Water Sallow.
S. aurita. Trailing Sallow.
S. Caprea. Great Sallow.
S. ambigua.
S. repens
S. fusca
S. prostrata
S. ascendens
S. incubacea } Dwarf Willows.
Populus tremula. Aspen.
Betula alba. Silver Birch.
***Fagus sylvatica.** Beech.
Quercus pedunculata. Oak.
Corylus Avellana. Hazel.
***Carpinus Betulus.** Hornbeam.
Paris quadrifolia. Herb Paris.
Tamus communis. Black Bryony.
Orchis mascula. Early purple Orchis.
***O. maculata.** Spotted Orchis. Damp spots.
Habenaria bifolia. Butterfly Orchis.
***Spiranthes autumnalis.** Ladies'-tresses.
***Neottia nidus-avis.** Bird's-nest Orchis. Among Beech leaves.
Listera ovata. Tway-blade.
Epipactis media. Helleborine.
Iris fœtidissima. Stinking Iris.
***Convallaria majalis.** Lily of the Valley.
***Polygonatum multiflorum.** Solomon's Seal.
***Ruscus aculeatus.** Butcher's Broom.
Scilla nutans. Bluebell.
Juncus effusus. Soft Rush.
J. conglomeratus
J. glaucus
J. diffusus
J. acutiflorus
J. lamprocarpus
J. squarrosus
J. bufonius } Rushes.
Luzula campestris
L. multiflora } Wood-rushes.
Alisma Plantago. Water Plantain.

Alisma ranunculoides.	Lesser Water Plantain.
Typha latifolia **T. angustifolia**	Reed-maces.
Sparganium simplex **S. neglectum**	Bur-weeds.
Arum maculatum.	Lords and ladies.
Lemna minor **L. gibba**	Duckweeds.
Potamogeton natans **P. polygonifolius** **P. heterophyllus**	Pondweeds.
Eleocharis palustris ***E. multicaulis** ***E. acicularis**	Spike-rushes.
Scirpus pauciflorus **S. fluitans** **S. setaceus**	Club-rushes.
***Eriophorum angustifolium** (?).	Cotton Grass. Bogs.
Carex pulicaris **C. disticha** **C. vulpina** **C. paniculata** **C. ovalis** **C. acuta** **C. pallescens** **C. panicea** **C. præcox** **C. pilulifera** **C. glauca** **C. flava** **C. binervis** **C. Pseudo-cyperus** **C. hirta** **C. paludosa** **C. riparia**	Sedges.
Phleum pratense.	Cat's-tail Grass.
Alopecurus pratensis.	Fox-tail Grass.
A. geniculatus.	Floating Fox-tail Grass.
Nardus stricta.	Mat-grass.
Phragmites communis.	Reed.
Calamagrostis Epigejos.	Wood Reed.
Agrostis canina **A. vulgaris** **A. pumila** **A. alba**	Bent-grasses.
Holcus lanatus **H. mollis**	Soft-grasses.
Aira cæspitosa **A. flexuosa** **A. caryophyllacea** **A. præcox**	Hair-grasses.
Avena flavescens.	
Arrhenatherum avenaceum.	
Triodia decumbens.	
Molinia cærulea.	Purple Melic-grass.
Poa annua **P. nemoralis**	Meadow-grass.

Poa trivialis. Meadow-grass.
Glyceria aquatica. Sweet-grass.
Cynosurus cristatus. Dog's-tail Grass.
Dactylis glomerata. Cock's-foot Grass.
Festuca myurus. Wall Fescue-grass.
F. ovina. Sheep's Fescue-grass.
Bromus asper, **B. sterilis**, **B. mollis**, **Brachypodium sylvaticum** } Brome-grasses.
Triticum repens. Couch-grass.
Hordeum murinum. Wall Barley.
Lolium perenne. Rye-grass.
Taxus baccata. Yew.
Juniperus communis. Juniper.
Lysimachia Nummularia. Creeping Jenny.

Among the FERNS found on the Forest the most abundant are *Pteris aquilina*, the common brake-fern, which grows wherever it is dry; *Scolopendrium vulgare*, the Hartstongue; *Lastrea filix-mas*, the Male fern; *Polypodium vulgare*, the Polypody which clothes the crowns of pollards. For further information about FERNS, see *Trans. Essex Field Club*, vol. i. p. 25; iii. pp. lxi. lxii.; ii. p. lxv.

EPPING FOREST

ITS MOSSES

"Beneath the birch with silver bark,
And boughs so pendulous and fair,
The brook falls scattered down the rock
And all is mossy there."

COLERIDGE.

THE majority of those who come to the Forest confine their visits to the summer months, leaving its glades comparatively deserted in what, to my mind, are the two most beautiful periods of the year—May, when the birches and beeches put on their tender green, and again when, in the late autumn, the same leaves turn to bronze and gold, contrasting with the lustre of green moss and the spotted beauties of the fungus family.

The mosses only begin to put forth their most brilliant greens and feathered growth when other vegetation is withered and dead, and it is in what are generally known as the dull months that many of them are in the most interesting stage to the naturalist and collector—viz. in fruit. These plants lend themselves more readily than any other to the art of mounting and preserving, as they retain their brilliancy of colour for many years, and their delicacy and variety of form are well shown on the mounting card. Mr. English has furnished me with the following list of the more important of the mosses which he has collected in the Forest, with their localities. I append it for the benefit of the numerous class of students who make these plants the object of, and the excuse for, winter rambles.

Sphagnum rubellum. Forming rosy tufts, on the margin of boggy places. By the side of Copped Hall Lodge Road.

Dicranella heteromalla. This little gem of early spring, with its hair-like foliage and glowing orange fruit, grows under oaks and by the side of old gravel-pits.

Dicranum scoparium. A pretty light green plant, with foliage slightly curved, in dense masses, as soft as a cushion.

Leucobryum glaucum. Glaucous green, quite a departure from the usual run of mosses, appearing like mounds rather than moss; rarely found in fruit. In and near Monk Wood.

Ceratodon purpureus. Most abundant by the side of the Theydon Road on old charred mounds where the heather has been burned.

Funaria hygrometrica. On charred places where fires have occurred.

Philinotis fontana. A curious forked plant; very local; grows in a bog by the Wake Arms, Theydon Road, and Loughton Road.

Bryum argenteum. A pretty silvery plant. On banks near Theydon.

Mnium undulatum. Moist places. Epping Lower Forest.

Mnium hornum. Sometimes called Swan's Neck Moss, from the curvature of the upper part of the fruit-stalk.

Aulacomnium palustre. Damp and boggy places.

Tetraphis pellucida. On old stumps and damp places. Close by Ambresbury Banks.

Polytrichum commune. In bogs.

Polytrichum juniperum. Gravel-pits by Wake Arms.

Thuidium tamariscinum. Very handsome, bright green, fern-like.

Eurynchium striatum. Epping Lower Forest.

Hypnum cuspidatum. Damp places. Bogs.

Hypnum schreberi. In masses amongst heath. Common.

Hypnum splendens. Often shaded with bronze. Very elegant, like the frond of a fern.

Hypnum triquetrum. This is the well-known "French Moss" sold at Covent Garden for decorations.

Lycopodium clavatum. Club moss, sometimes called Stag's-horn Moss. Between Monk Wood and Cowper's Camp.

The following mosses have also been gathered in the Forest, but it is probably far from an exhaustive list of those inhabiting the locality—

Atrichum undulatum.
Aulacomnion androgynum.
A. palustre.
Bartramia fontana.
B. pomiformis.
Bryum argenteum.
B. cæspiticium.
B. capillare.
Bryum carneum.
B. pseudotriquetrum.
Campylopus flexuosus.
Ceratodon purpureus.
Dicranum cerviculatum.
D. heteromallum.
D. fuscescens.
D. scoparium.

Fissidens bryoides.
F. taxifolius.
Fontinalis antipyretica.
Funaria hygrometrica.
Grimmia pulvinata.
Hypnum aduncum.
H. albicans.
H. confertum.
H. cupressiforme.
H. cuspidatum.
H. denticulatum.
H. prælongum.
H. purum.
H. riparium.
H. ruscifolium.
H. rutabulum.
H. Schreberi.
H. serpens.
H. splendens.
H. squarrosum.
H. striatum.
H. tamariscinum.
H. triquetrum.
H. undulatum.
H. velutinum.
Isothecium myosuroides.
I. myurum.
Leskea sericea.
Leucobryum glaucum.
Mnium hornum.
M. punctatum.
M. undulatum.
Neckera complanata.
Orthotrichum affine.
Phascum cuspidatum.
P. nitidum.
P. serratum.
P. subulatum.
Physcomitrium pyriforme.
Pogonatum aloides.
P. nanum.
Polytrichum commune.
P. juniperinum.
P. piliferum.
Pottia truncata.
Sphagnum acutifolium.
S. contortum.
S. cymbifolium.
Tortula lævipila.
T. muralis.
T. subulata.
T. unguiculata.
Weissia cirrhata.
W. controversa.
Zygodon Forsteri (gathered by Mr. Holmes).
Z. viridissimus.

EPPING FOREST

ITS FUNGI

"You demi-puppets, that
By midnight do the green-sour ringlets make
Whereof the ewe not bites; and you, whose pastime
Is to make midnight mushrooms."—*The Tempest.*

I DO not know that it is anything to be proud of, but, as a matter of fact, Epping Forest is blessed with an abundance of toadstools beyond any other district within easy reach of London. The surface of the ground and the stems of partially decayed trees are studded, in the autumn months, with a great variety of forms of fungi, some remarkable for the brilliance of their colours, others noteworthy for their curious forms. In size they vary from huge gnarled specimens a foot in diameter to the minutest thread-like mildew. Some are fashioned like parasols, others are concave and resemble vases or goblets. Some are circular, some columnar, and others have no defined shape at all. The colours comprise black, pearly white, purple, orange, yellow, and the most brilliant scarlet. Some are smooth and satiny in appearance, others studded with scales and bosses of darker colour. Their characteristics change to an extraordinary extent with the different stages of their growth. In duration they vary greatly. "Many spring up in a night and dissolve next day, or at least at the first shower of rain. Some are so delicate that they tremble in the hand when gathered, and, as in the case of *Coprinus radiatus*, melt away if breathed upon. Others, growing on old trees, are perennial, and almost as hard as wood." Their simplest forms consist of minute articulated filaments. Their highest developments are the brilliant domes and minarets which are called *Agarics*, and which jewel the mossy banks in September and October. They are sought by many naturalists, not only for their intrinsic beauty, but for their varied structure, and the interesting problems which they present to the botanist and the microscopist. The bolder spirits try experiments in their edible qualities, but this cannot be done with impunity, except by those who have acquired intimate knowledge of the varieties.

The Forest Field Club annually devotes a day to a "fungus foray," and generally secures some rare and beautiful specimens. Their habitats vary greatly, but all derive their nourishment from decaying animal or vegetable matter. It is difficult to localise the species, as, having found a rare kind in a certain spot once is no guarantee that it will reappear there the following year, and it may even be many years before the conditions favourable to its production will repeat themselves. The vegetative system is the mycelium, or rooting fibrils, which extend beneath the surface of the soil, and may exist for years in that condition, without producing the full development in stem, branches, or cap, which produce the fructification that is analogous to the flower or fruit of a flowering plant. I am indebted to Dr. M. C. Cooke, one of the greatest authorities on the subject, for the following very interesting notes on the various species most likely to be found in the Forest. He writes :—

As not less than six hundred species of the larger fungi have been found within the limits of the Forest, it would only be embarrassing to give a record of all their names. We shall rest content with the enumeration of the most common, or most remarkable.

EDIBLE FUNGI

Amanita rubescens. Growing on the ground to the height of four or six inches, with a cap of four inches in diameter, of a purplish brown colour, sprinkled with small irregular pale warts. Common all over the Forest from July to October.

Lepiota procera. Parasol mushroom. Stem six or eight inches long, bulbous at the base. Cap often six inches across, with a strong boss in the centre. Silky gray, clad with brown scales. In open places, High Beach, Fairmead, etc.

Lepiota rachodes. A similar species, without such a distinct boss, and with more persistent scales, is less common.

Armillaria mellea. Very common in large clusters around dead stumps. Cap honey colour or pale brown, rather scaly, two to three inches broad. Largely eaten in Austria, but not approved in this country.

Armillaria mucida. Ivory white, covered with a thick coat of slimy gluten, grows in tufts on dead beech trunks; of a delicate flavour when cooked.

Tricholoma nuda. Entirely lilac or amethyst colour, cap from two to four inches, quite smooth. Growing amongst dead leaves. Chingford, Loughton, and High Beach.

Tricholoma personata. Blewits. Not so common as the last. The cap about three inches, of a dirty white; the stem rather short and thick, and usually tinged with purple. Odour strong. In open places.

Clitocybe nebularis. Cap four inches or more in diameter, mouse colour. Stem thickened downwards, four to six inches long; gills running down the stem. Odour strong and mealy. Grows amongst dead leaves.

Clitocybe infundibuliformis. The cap about two inches in diameter, depressed in the centre, so as to be funnel-shaped; gills white and thin, running down the slender stem. Amongst dead leaves everywhere. When dry almost of the colour of wash-leather.

Clitocybe fragrans. Smaller than the last. Cap not more than one inch, and not so deeply depressed. Whole plant fragrant, like hay, or the sweet woodruff, retaining fragrance when dry.

Clitocybe baccata. One of the most common fungi in woods. Usually of a dull, brick-red colour, including the gills, which are thick. The cap an inch in diameter, and convex; stem two to three inches long and slender. Harmless, but not very delicate eating.

Clitocybe odora. This is also a fragrant species, with a faint odour of anise. The cap, about two inches in diameter, is of a dull glaucous green and nearly flat, with a slender stem. It grows amongst dead leaves, and has been found in Lower Epping Forest.

Collybia fusipes. Entirely of a reddish brown colour, growing in tufts on rotten wood. Cap from two to three inches; stem rooting, thickest in the middle, tapering towards each end. Hawkwood, Theydon Bois, etc.

Pleurotus ostreatus. Oyster mushroom. Grows in dense tufts on old stumps, with the caps overlapping like tiles, mostly of a mouse colour or pale gray, with very white gills. Stem short and oblique. A tuft will sometimes measure nearly a foot across. Lower Forest, High Beach, etc.

Pleurotus ulmarius. Sometimes singly and sometimes two or three together, on dead elms. Creamy white or pale tan colour. Cap from two or three to six or eight inches in diameter. Thick and fleshy, with a mealy odour. Near Chingford and Loughton.

All the foregoing species have white spores, which fall upon the ground beneath, when mature, like chalk powder.

Fungi which have pink or salmon-coloured spores are rarely edible, and often poisonous.

Clitopilus prunulus. Plum mushroom.

Clitopilus orcella. Sweetbread mushroom.

These are two species with pink spores, and are nearly of the same size and appearance. The cap is almost flat at first, then concave; ivory white with a satiny lustre, and rather sticky when moist. About two inches in diameter, with a short stem and pale purplish pink gills. Odour strong and mealy. Both found in the Lower Forest, as well as around High Beach.

Of edible species with purple-brown spores the most useful are:

Agaricus campestris. The common mushroom, which should be very well known, and has at first pink gills, afterwards dark purple-brown. Grows in open pastures.

Agaricus arvensis. The horse mushroom, generally of a larger size, with gills at first of a dirty white (never pink), and afterwards becoming nearly black. Lower Forest.

Coprinus comatus. "Shaggy caps." Grows in tufts on the ground. The cap three or four inches long, and half as thick, is covered with shaggy fibres, wholly dirty white. The gills are whitish until the cap begins to expand, when they become black, and at length melt away in drops, like treacle. Common on waste ground.

Coprinus atramentarius. "Inky mushroom." Grows in large tufts, mostly near rotten wood, posts, etc. Cap smooth, smoky gray, not so cylindrical as the last, sometimes larger. When expanded the gills melt and drip away like ink. Edible when young.

Hygrophorus eburneus. Pure white. Cap less than two inches, stem longer. Slimy when moist. Found on the elevated sides of Monk Wood.

Hygrophorus virgineus. Ivory white, sticky when moist, cap one or two inches broad, becoming flat; gills broad, and distant from each other, running down the short stem. Open spaces and glades, King's Oak to Wake Arms.

Paxillus involutus. Large and very common under trees. Cap four to six inches broad, with the edge at first rolled inwards. Gills close together running down the stem. Wholly dark tan-colour, dirty brown when moist. Edge of cap woolly.

Cantharellus cibarius. "The chantarelle." Whole plant of an egg-yellow colour, two or three inches high, two inches broad. Gills very thick and shallow, like veins, running down the short stem. Odour of apricots. Not uncommon under trees.

Marasmius oreades. "Fairy ring champignon." Only found in the open and not very common. Cap about an inch broad, pale tan-colour. Gills broad, and wide apart, nearly white. Stem slender. The whole fungus dry and tough. High Beach and Lower Forest.

Boletus edulis. The under side of the pileus is not furnished with gills, but closely packed pores like pin holes. Cap three or four inches or more across, convex, smooth like a bun, and brown, under side greenish yellow. Stem thick, often two inches below, solid, paler than the cap, the upper part netted with delicate veins. Common everywhere.

Boletus scaber. Differs from the last in the stem being longer and more slender, covered with rough, dark, projecting points; under side of the cap dirty white or grayish, with a hollow round the stem. Common under trees.

Fistulina hepatica. Vegetable beefsteak. Resembles a raw ox-tongue in form and colour, and when cut in section the flesh is mottled somewhat like beetroot. For several years it grew upon an old oak outside Fairmead Lodge, and has been obtained from several other localities in the Forest.

Hydnum repandum. "Hedgehog mushroom." The under side of the cap, instead of gills, is furnished with spines or teeth. Whole plant of a delicate pinkish cream colour, very irregular, two to four inches across. Stem short. Hot to the taste. Common under trees everywhere in favourable seasons.

Craterellus cornucopioides. "Horn of plenty." About two or three inches high, trumpet-shaped, or long funnel-shaped, brown within, and dark gray, almost black, externally. Substance thin and tough. Usually growing in company. Sometimes common on the slopes of Monk Wood.

Clavaria. Several species are found in different parts of the forest growing amongst grass. Some are simple and club-shaped, single or in clusters; others are branched, and mostly either white or yellow. All are comparatively small for edible purposes.

These are not all the edible fungi which have, from time to time, been found in the Forest, but this selection includes those which are most likely to be met with during an autumnal stroll.

POISONOUS FUNGI

We do not believe that there are so many kinds of fungi which are really poisonous as popular opinion has supposed for so long, although there are a few about which there should be no mistake.

Amanita muscaria. Fly agaric. A splendid species six or eight inches in height, with a cap three to six inches across, of a rich orange scarlet, sprinkled with white angular warts. Under birch trees. Plentiful throughout the shady parts.

Amanita phalloidea. Stinking amanita. Not so large as the above, with a greenish yellow cap to which adhere patches of the cuticle of the volva. The stem has a bulbous base rising out of the remains of a sheath or volva. Under trees in all shady places. Stinking in decay.

Amanita pantherina. Resembling in appearance and size the Amanita rubescens, but duller in colour, with no tinge of red, and not so common. If not truly poisonous, it produces disagreeable effects. Under trees, High Beach and elsewhere.

Entoloma sinuatus. A large species with the cap four inches across, of a pale mouse colour. The stem, six inches in length, often contorted. The gills broad, and of a salmon colour. It grows in company in shady places near Loughton and Chingford.

Hebeloma fastibilis. The cap two or three inches in diameter, resembling a cracknel biscuit; of a rather strong disagreeable odour. About the size of an ordinary mushroom. Not common in the Lower Forest, and on the Chingford side.

Stropharia œruginosus. Often found amongst grass and dead leaves, sometimes singly and sometimes in small tufts. The cap averages two inches in diameter, is of a deep verdigris green and very slimy. The gills are dark brown. Throughout the shady parts of the Forest.

Stropharia stercorarius.
Stropharia semiglobatus.

These are two similar species found upon dung. The cap is convex, and less than an inch across, with broad smoky gills. The stem is four inches long or more, about the thickness of wheat straw, and the whole plant is glutinous.

Hypholoma fascicularis. Grows in dense clumps about old stumps. Its general colour is yellow and the top of the cap often reddish; the gills are a dingy olive. It is intensely bitter to the taste, and too nauseous to be eaten.

Psilocybe semilanceatus. This little agaric is very common in open places amongst grass. The pileus about half an inch across and conical, of a pale ochre. The stem is long, thin, and flexuous. Several instances are known in which death has accrued to children through eating them.

Hygrophorus conicus. Grows singly amongst grass, and is of a dingy yellow colour, soon turning black. The cap is conical and pointed, about an inch broad and high, with a slender fragile stem.

Hygrophorus psittacinus. Grows also amongst grass in open places, and similar in form and dimensions to the foregoing, but of a bright pale green colour, turning yellowish but not black. Both are covered in moist weather with a sticky gluten.

Lactarius torminosus. This is one of a genus or group which, when broken, exudes from all parts a white or coloured milk. The cap is about three inches across, depressed in the centre, with the margin woolly and curved inwards. Dull orange-red with darker bands. High Beach and Lower Epping Forest.

Lactarius vellereus. Another of the milky agarics, found everywhere in the most wooded parts. Wholly of a dull white. The cap six to ten inches in diameter, deeply depressed, almost funnel-shaped, and covered with a dense fine woolliness. The gills are broad, and run down into the short thick stem. Milk-white and peppery.

Lactarius piperatus. Similar to the above, but not so large. The top of the cap quite smooth; the gills narrower and twice as numerous. Milk-white and biting. Always considered to be poisonous in this country, but sometimes eaten in America.

Russula emetica. The cap, about two inches across, is rosy red, smooth, and shining. The stem as well as the gills, of a pure white. The entire fungus is very fragile. The cuticle of the cap peels off readily and shows pink flesh beneath. Common to the whole Forest, but not numerous.

Russula fellea. Similar in shape and size to the above, but wholly of the colour of wheat straw, including the gills and the flesh. It is intensely bitter. At High Beach, Fairmead, Theydon, and towards Chingford.

Russula fœtens. Stinking Russule. Often with the cap six inches across, dirty amber yellow, soon depressed in the centre, margin bent over, and marked with parallel furrows. Stem short. Mostly of a strong stinking odour, but occasionally found without smell. Scattered over the entire Forest, in woody places.

Marasmius personatus. Very common everywhere amongst dead leaves. Cap convex, two inches across, pale purplish brown, quite dry and tough; gills brown, but the spores white. Stem three inches long, thick as a clay-pipe. The lower half clothed with a dense pale-yellowish down.

Boletus satanas. Cap convex, like half a globe, from four to ten inches across, dirty white, rather sticky. Under surface flat, bright livid crimson. Stem red and yellow, two inches thick. When cut, bruised, or broken, quickly turning in colour to a deep indigo-blue. Only recorded in the Lower Forest.

Boletus luridus. Cap similar in form, semi-globose, but not more than four inches across, olive-brown and velvety. Under surface lurid red. Stem thick in proportion, orange above and red or brown below sprinkled with dots. Changing when wounded to indigo-blue. High Beach and Fairmead, usually.

Boletus piperatus. Smaller and more slender, cap flattened convex, about two inches across, tan-coloured with a tinge of red. Under surface with large angular pores. Stem slender. Flesh yellow. Taste very hot. High Beach and Lower Forest.

FUNGI OTHERWISE OF INTEREST

Amanita vaginata. On the ground everywhere. Cap at first bell-shaped, then open and convex, rather viscid; mouse-gray, brown or sometimes white. Fragments of membrane of the volva sticking to the cap. Margin marked with close-set, fine, parallel lines. Stem six inches long, white, without ring. Base standing in the remains of a membrane—like sheath or volva.

Lepiota granulosa. A small species found amongst short grass. Cap less than an inch across, convex, warm tan-colour, covered with small powdery granules. Gills white. Stem slender, coloured below and powdery.

Tricholoma saponacea. Gray-capped fungus smelling of soap. Cap two to three inches in diameter, smooth and dry. Stem three inches long, solid, sometimes scaly. Monk Wood, High Beach, etc.

Tricholoma terrea. Mouse-gray fungus, many growing together on naked ground. Cap two to three inches across, clothed everywhere with closely-pressed hairs. Stem rather short. Common over all the Forest.

Collybia radicata. Solitary. Cap from two inches in diameter, soon flattened, pale brown, with radiating ribs or veins. Gills very broad, and distant from each other. Stem slender, rigid, very long, six inches or more above the ground, and three inches of rooting base, immersed in the soil. Common over the Forest area.

Collybia butyracea. One of the commonest fungi in the Forest. Cap two to three inches, with a broad hump in the centre, dark when moist, pallid when dry, flesh thin. Stem purplish, very hollow, easily flattened by the fingers.

Collybia dryophila. Another very common species, with a pale, flat, ochrey-white cap, from an inch in diameter, with thin, narrow white gills. Stem slender as a straw, reddish brown. Grows on dead leaves.

Mycena galericulata. Growing in dense tufts on stumps. Cap conical, gray, hardly an inch across. Gills white or tinged with pink. Stem rigid, smooth, darker than the cap, four to six inches long.

Mycena sanguinolenta. One of half a dozen species which contain milk in the stem. This small one with a cap half an inch across and a long thready stem, is found amongst dead leaves. When the stem is broken a drop of blood-red milk oozes from the wound. High Beach, Fairmead.

Nolanea pascua. A little dark gray fungus found commonly in pastures, amongst short grass. The cap is obtusely conical, one inch broad and high, with a silky appearance. The gills are pink. Found at Loughton and on Chingford plain.

Pholiota spectabilis. Is a splendid golden-coloured agaric, which grows in small tufts out of rotten trunks. The cap is three or four inches across, the flesh thick and sulphur yellow. Thick stem four to six inches long, with a large ring, or collar, near the top. Gills and spores rusty. Over the Forest area.

Psilocybe spadicea. Forms immense clumps about old stumps. Cap convex, bay brown when moist, one to one and a half inches in diameter, paler when dry. Stem white. Gills and spores dark purple brown. Over the whole Forest.

Paneolus fimiputris. Cap conical, not an inch high or broad, thin, dark, viscid when moist, shining when dry, with a rigid erect pallid stem. Gills and spores black. Growing on dung. Southern parts of the Forest.

Coprinus micaceus. Cap oval, then bell-shaped, sprinkled with shining atoms; golden brown, marked at the margin with close parallel lines. Gills and spores black, soon melting. Common on stumps.

Cortinarius elatior. Cap oval, and then expanded, very viscid, as well as the stem; brownish, shining when dry, two to four inches. Stem long, scarcely an inch thick, attenuated both ways; white, tinged with violet. Gills rust colour. Under beech trees, High Beach and several other places. Common.

Cortinarius cinnabarinus. About one inch across, blood-red. Stem slender, two inches long, of the same colour. Amongst dead leaves. Monk Wood, High Beach.

Russula cyanoxantha. Cap two to three inches, convex, smooth, rather variable in colour, bluish towards the margin. Gills and stem white. Mild to the taste. High Beach, Loughton, Chingford.

Russula nigricans. This will be recognised most often in its old state, when it turns black all over, and may be seen in all the woody parts of the Forest, rotting on the ground. Cap from four to six inches across, and the gills very thick and broad.

Nyctalis asterophora is a small, pale, mealy-looking fungus about the size of a pea, which grows in the autumn, as a parasite upon the foregoing, in a state of decay.

Lenzites betulina. A leathery, shell-shaped fungus on old stumps, about two inches across, without stem, attached by one side. The top is pale and hairy, whilst the under side has tough leathery gills.

Boletus chrysenteron. The cap is from two to three inches across, a little convex and velvety, mostly olive-brown, cracking and exposing pink flesh beneath the cuticle. Stem three inches long, often dark red, and the pores greenish yellow. Common everywhere.

Polyporus adustus. Has the pileus, or cap, shell-shaped, an inch or two across, attached along one side to old stumps. The upper side is of a smoky gray, but pale, with a darker margin. The under side is porous and silvery gray.

Polystictus versicolor. Similar to the above in form, but the pileus is very variable in colour, with concentric zones or bands of a darker colour. The porous under side is of a snowy white. Both are found on dead stumps all over the Forest.

Stereum hirsutum. Cannot fail to be noticed on almost

every dead stump. The irregular-shaped pileus is attached along the black edge, whilst the upper surface is olive and hairy, with a golden-yellow edge. The under surface is smooth, without gills or pores, and golden yellow.

Stereum purpureum. Similar to the last but smaller, and the under surface of some tint of purple, instead of yellow.

Tremella foliacea. Soft and jelly-like masses, as large as a walnut, or small apple, of a brownish colour, on rotten trunks. Substance yielding to the touch, contorted and twisted, shaking like jelly when removed. Lower Forest, Fairmead, Theydon Bois.

Exidia glandulosa. Smaller and black, with scattered spiny projections. Firmer in substance, and found growing on fallen oak branches. Loughton side.

Phallus impudicus. The stinkhorn, well known from its strong odour and singular erect form, will be detected in all parts of the Forest by the smell, even if it should not be seen.

Hundreds of other kinds of fungi are to be found in different parts of the Forest, which it would be impossible to include in a book of this size. Coloured figures of many hundreds of them may be seen in the Forest museum.

There are, besides those enumerated, hundreds of species of fungi to be found in the Forest district, which are only of interest to those who study the subject specially, the majority being only just visible to the naked eye. Of these the most important are the rusts, smuts, and mildews, parasitic on growing plants,[1] more or less destructive to vegetation, and injurious to forest trees.

In addition are the minute species which flourish on dead vegetation, decaying leaves, fallen twigs and branches, old stems, and prostrate trunks. All these are constantly at work, destroying the vestiges of an old vegetation, and preparing a vegetable humus, for the growth of a new generation.

The pretty cups of Peziza aurantia, from an inch to two or three inches in diameter, may often be seen growing in company on the damp ground. These may be taken as the type of about sixty species recorded for the Forest,[2] some of them so minute as only to be just visible to the naked eye.

To these might be added fully sixty species of minute moulds, which have been but little sought after, and might probably soon exhibit a list of double that number, all of which require the use of the microscope.

Those who are interested in those extraordinary organisms called Myxogasters, which in one stage of their existence resemble animals, and in another seem to be plants, would find a good field for their researches in the Forest, which, according to Mr. Lister, is a happy hunting-ground.

[1] A list of some sixty-eight Forest species was given in the *Essex Naturalist* of September 1887.
[2] *Essex Naturalist* for 1888.

EPPING FOREST

A SKETCH OF THE GEOLOGY OF EPPING FOREST[1]

By T. V. Holmes, Esq., F.G.S.

A glance at a geological map of South-eastern England shows that London stands in the midst of a district of Tertiary clays and sands. These occupy the surface of a large tract of country, which, widening eastward till it includes Ipswich and Canterbury, becomes narrower westward and dies away around Newbury and Marlborough. Immediately outside this Tertiary area appears a belt of country of very variable breadth covered by the well-known soft limestone known as the Chalk, which underlies the Tertiary district, and comes out to the surface in Kent, Surrey, and Hampshire southward, and northward in Suffolk, North-western Essex, Hertfordshire, and other counties. In Essex, south of the Tertiary area, Chalk appears only in the small district between Purfleet and East Tilbury; north of it Chalk may be seen north-west of a line drawn from Ware through Bishop Stortford to Sudbury.

Thus in Epping Forest only Tertiary, and rocks of still later date, appear, the oldest being that known as the London Clay. The following section gives the nature and thicknesses of the strata found beneath the London Clay in a deep boring at the Great Eastern Railway Station, Loughton (*Trans. Essex Field Club*, vol. iv. pt. ii. p. 158):—

[1] Those requiring fuller information may be referred to Mr. W. Whitaker's *Memoir* on "The Geology of London, and of part of the Thames Valley." (Stanford.) 2 vols.

		ft.	in.
	London Clay (surface) .	167	0
Reading Beds	Sand	26	0
	Clay and Stones . .	10	0
	Thanet Sand . .	40	0
	Chalk . .	648	6
		891	6

The London Clay forms probably about four-fifths of the surface of Epping Forest, the remaining fifth consisting of small patches of sand, boulder clay, and gravel which lie here and there upon the London Clay. They are of various geological dates, and may be seen at various heights above sea-level. Though they are all comparatively thin, and geologically obvious mainly from a more flattened contour of the ground than results from uncovered London Clay, they give rise to a considerable diversity in the aspect of the Forest from the space they afford for those trees and plants which flourish on a sandy, not on a clayey soil.

The London Clay when seen in a fresh cutting is of a slaty-blue colour, but soon becomes brown on exposure to the air. In a large section the layers of concretionary clayey limestone, with septa of crystalline carbonate of lime, known as Septaria, are usually to be observed. The greatest thickness attained by the London Clay is in Southern Essex, and is about 500 feet. At Dagnams, near Romford, it was found to be 400 feet thick, and it is probably equally thick at High Beach. It appears to have been deposited in a comparatively shallow sea near the mouth of a large estuary, many of its fossils being vegetable remains from the nearest land. But crustacea and mollusca also abound, the most common molluscs belonging to the great group Gasteropoda, of which the whelk and snail are familiar examples. Though there are plenty of fossils in the London Clay, they are not likely to be easily detected in the sections of Epping Forest, though conspicuous on the foreshore beneath a cliff of London Clay, as at

Herne Bay and Sheppey. They indicate a climate decidedly warmer than the Essex of to-day.

The London Clay extends far westward as well as eastward of Epping Forest, underlying the river deposits of the Lea as well as those of the Roding or the Crouch. Within the area of the Forest where it does not form the surface it may be met with at a depth seldom exceeding 20 to 30 feet. Its uppermost beds are usually loamy or even sandy. They may be seen in Oak Hill pit, about a mile west of Theydon Bois Railway Station. It is sometimes difficult to distinguish between the uppermost beds of the London Clay and the lowest of the Bagshot series, which are next in age.

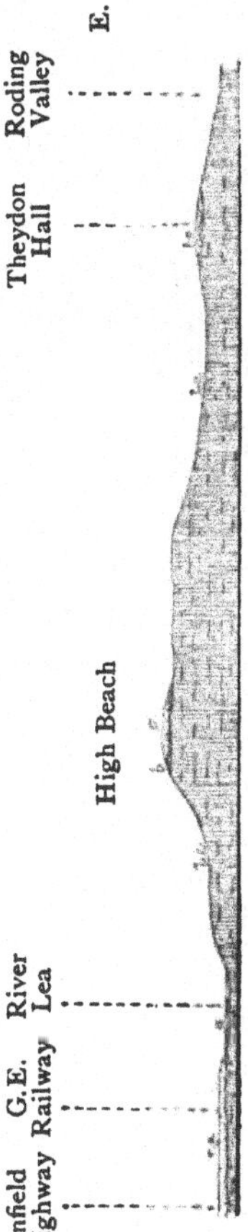

Fig. 1.—Diagram Section from Enfield Highway (W.) across High Beach to the Valley of the Roding east of Theydon Hall (E). Length of Section, 8 miles. *l.c.*, London Clay. *b.*, Bagshot Beds. *g.*, the "Pebble Gravel" of Geol. Survey Map. *b.c.*, Boulder Clay. *r.g.*, River Gravel. *a.*, Alluvium.

The Bagshot Beds, though they occupy a considerable area around Bagshot Heath and Aldershot, form but little of the surface of Essex, though small, scattered, outlying patches of them appear at Rayleigh, near Brentwood, and between Epping and Ongar. They consist of sand with partings of clay. Within the present area of Epping Forest there is but one small outlier, at High Beach, a section in which was visible a few years ago when the reservoir was being made. (See *Essex Naturalist*, vol. i. p. 107.)

The London Clay and the Bagshot Beds both

belong to the Eocene series. In other words, the Bagshot Beds were deposited above the London Clay in the sea in which the latter had been laid down, no interval having existed during which upheaval of the area, and, consequently, more or less erosion of the London Clay, occurred. The change was simply in the nature of the sediment, which became more sandy. Thus it becomes evident that the London Clay, was not reduced in thickness during the deposition of the Lower Bagshot Beds, while it is certain that a similar remark cannot be made in the case of any other of the deposits which overlie the London Clay. Consequently, only where the London Clay is capped by Bagshot Beds can its full thickness be ascertained. And the only spot within the Forest area where this occurs is High Beach.

There is a very considerable interval, as regards age, between the London Clay and Bagshot Beds and the gravel which is found here and there capping the ground in the higher parts of the Forest, sometimes resting, as at High Beach, on Bagshot Beds, sometimes on London Clay. It may be seen at High Beach and Jack's Hill within the area of the present Forest, and at Gaynes Park and Coopersale Common just outside its limits. The late Sir Joseph Prestwich classed this gravel, which has a very wide distribution, as "Westleton Shingle." But those who wish to know what is meant by that term must be referred to his important paper on the "Westleton Beds" in *Quart. Journ. Geol. Soc. for* 1890, pp. 84-181. He gives the approximate composition of the gravel at Jack's Hill as:—

1. Flint pebbles	50	per cent
2. White quartz pebbles . . .	15	,,
3. Subangular fragments of flint .	20	,,
4. Subangular fragments of chert and ragstone	10	,,
5. Pebbles of Lydian stone, etc. .	5	,,
	100	

While at Coopersale Common or Gaynes Park he states it to be:—

1.	Flint pebbles	56 per cent
2.	White quartz pebbles	20 ,,
3.	Subangular fragments of flint	9 ,,
4.	Subangular fragments of chert and ragstone	12 ,,
5.	Pebbles of Lydian stone, etc.	3 ,,
		100

Small patches of this gravel may be seen at Earl's Path (the road from Loughton to High Beach), and also along the road between the Robin Hood Inn and the Wake Arms. Though, as before remarked, of much later date than the Bagshot Beds, the Westleton Shingle, though post-Pliocene, is yet pre-Glacial in age. Prestwich notes, *inter alia*, that it contains pebbles of southern origin unknown in Glacial Deposits.

The beds belonging to the Glacial Period consist, in this district, of sand and gravel underlying boulder clay. The sand and gravel varies very much in composition and in general character and aspect, while the Boulder Clay, commonly known as the "Chalky Boulder Clay," from the numerous pebbles of chalk it contains, is singularly invariable in appearance, whether seen in Norfolk, Suffolk, Essex, or Hertfordshire. In composition the Glacial Gravel consist chiefly of flint, subangular or rounded; quartz and quartzite, flint greatly predominating. Sometimes rolled fossils from the Lias and Oolite have been found in it. The Chalky Boulder Clay is easily identified wherever there is a section having a depth of 5 or 6 feet or more. But it often happens that the foot or two next the surface is so much altered by the dissolving away of the chalk pebbles and other calcareous matter through the agency of rain-water, as to be scarcely recognisable.

Patches of gravel of considerable size, with no covering of Boulder Clay, but which seem on general grounds to be, in all probability, of Glacial

Age, exist at Loughton, Buckhurst Hill, and Woodford. And the small patch at Queen Elizabeth's Hunting Lodge, with two others between that building and Fairmead Lodge, may possibly be of the same period.

The Boulder Clay, though it may be seen east of the Roding as far south as Chigwell, and west of the Lea at Finchley, has not hitherto been detected within the present Forest except at the extreme north, beyond Epping, where it forms most of the surface of "The Lower Forest." There are also two smaller patches close to the town of Epping. Possibly some future section may reveal it in, or on, the gravel of Woodford, Buckhurst Hill, or Loughton. It may be described as a confused mass of rocks of various ages in a matrix of bluish-gray or buff-coloured clay. The chief ingredients are chalk and flint, then come rocks and fossils belonging to the Lias and Oolite formations. *Belemnites* and *Gryphæas* are very abundant. The constituents of the Boulder Clay have evidently been derived from the north, not from the south, but it is impossible here to discuss the various theories held by geologists as to the particular form of ice-action, whether ice-sheet, icebergs, or coast-ice, which has produced the Boulder Clay of the Eastern Counties.

In the most southerly portion of the Forest, about Leyton and Leytonstone, Snaresbrook and Wanstead, the surface is largely composed of gravel of lower level and later date, belonging to the present systems of the valleys of the Thames and Lea. It was deposited, however, when those rivers flowed at a somewhat higher level than they now do, and having more fall, brought down coarser material. And in the alluvium or marsh deposits of the Lea and Roding we have the lowest in level and latest in date of all the geological formations in, or bordering on, the Forest district. Sections in these beds have been common in the neighbourhood of Walthamstow from time to time during the last thirty years, the East London Waterworks Company having made many additional reservoirs in them during that period. With much variation in details, these sections usually disclose clayey loam at the surface, then more or less peat and shell-marl, then sand and gravel at the bottom, of an excavation 7 or 8 feet deep. Among the mammalian remains found in these deposits in 1868, Dr. H. Woodward (*Trans. Essex Field Club*, vol. iii. pp. 1-29) mentions those of the wolf, fox, beaver, horse, wild-boar, reindeer, gigantic ox (*Bos primigenius*), and elk.

In short, the London Clay underlies all the other formations mentioned from the Bagshot outlier at High Beach to the recent alluvium of the Lea and Roding. At one time Bagshot Beds covered the whole of the area of Epping Forest and extended far beyond it to the east and west. Then much later, in post-Pliocene times, came the deposition of the Westleton Shingle, and, somewhat later still, of the Glacial Gravel and Boulder Clay. After the close of the Glacial Period the present river systems came into being, and the Lea, Roding, and Thames cut their way through the deposits already mentioned, deepening and widening their valleys as they did so, and leaving behind them the various beds of gravel, peaty matter and loam, which are the most recent of all.

While much planing down of the then existing surface must have taken place through the action of the sea, during the deposition of the Westleton Shingle, and through the agency of ice in some form during that of the deposits of the Glacial Period, the present scenery of Epping Forest can scarcely be said to owe anything to those distant and remote influences. For the sculpturing of the surface which has given us the present distribution of hill and valley is due to the action of rain and rivers much more recently, to agencies still at work, and which may be said to have originated with the present river system in post-Glacial times.

EPPING FOREST

ITS MANAGEMENT[1]

PUBLIC attention has of late years been often turned to the management of Epping Forest, and those responsible for it have been subject to criticism which has at times been neither just nor well informed. I do not complain of this publicity. On the contrary, I welcome every sign of intelligent interest in this public heritage. The agitation has, moreover, its good side. It has at least compelled us to justify to ourselves every step which we proposed to take, and to redouble our vigilance.

I propose in this chapter to lay down some of the leading rules that should, in my opinion, govern the management of the woodland in Epping Forest. It would be out of place to go too minutely into detail, but I will propound what I consider the most important general principles that should guide us. Without pretending to exhaust the subject I will name four—(1) Variety; (2) Preservation of natural features; (3) Restoration of a natural aspect where this has been interfered with in the past; (4) Reproduction with a view to the future.

(1) VARIETY.—We have been peculiarly favoured in Epping Forest by the fact that the various manors of the district have been under distinct management in the past. This, combined with great diversity of aspect, soil, and degrees of moisture, has given us more contrasts of vegeta-

[1] This chapter is an expansion of an address delivered at a meeting of the Essex Field Club, 21st October 1896.

tion than can be found in any other English forest. Our aim should be to preserve, and emphasise, this variety, which constitutes one of the greatest charms of our woodland.

Let me now endeavour to point out the special features which characterise the most important blocks of woodland, and, so far as I can, the reasons for the differences which we find, as well as the policy which I suggest should prevail in the future.

Beginning at the north end, the Lower Forest, beyond the town of Epping, contains the finest specimens which we possess of pollard hornbeams, with splendid buttressed stems. The naturally heavy and wet soil seems to favour this species. The individual specimens have always had more room here than is the case elsewhere. The purpose here has been to select the finest trees and groups, and give them a chance of developing still further, till in the future they may even vie with the splendid groups of old pollard hornbeams at Hatfield. The process, if not complete, is so forward that very moderate thinning will suffice in future. Along with the hornbeams are a good many unpollarded oaks and a few ashes.

Coming south of Epping, we find in Epping Thicks a far drier soil, naturally drained by the deep little valleys which cross it. These not only furnish in themselves a new and very beautiful feature, but favour, on the dry ridges, a splendid undergrowth of holly in conjunction with the finest beeches which the Forest can show. A large number of inferior hornbeam pollards have been taken out, so that the beeches are now seen in majestic groups. Below the wide green ride there are some well-grown oaks, and here also a few more of the paltry pollards might well be removed. When this is done, this beautiful wood will go on improving without further interference for many years. There are here a few unpollarded hornbeams of large size. These are a great rarity in Epping Forest.

On the opposite side of the valley we find a

totally different character. A part of Theydon Manor has been entirely cleared a generation or two back, and has developed a fresh young growth of holly and birch, with a surface-growth of heather. In another part of the same Manor the trees have been periodically cut down to the stool, leading to a bunchy growth which is not found in other manors. These bunches have become very "sticky" from overcrowding. I hope that a proportion of them will be removed, so that the remainder may put out laterals. There are some nice young spear oaks, and here and there a grove of birches has taken possession.

Crossing the Theydon Road we enter Theydon High Wood, which is distinct from any other part of the Woodland, and resembles the German beech forests, the smooth branchless stems standing very thickly. It would appear that in this wood some former Lord of the Manor had eradicated every other kind except a few small oaks. These were pollarded, whereas the beeches are "spears." Where some of these beeches have had room to put out laterals we have given them more room. Otherwise the foliage would all be confined to one plane, 40 feet from the ground, which gives a monotonous effect. Here a rich carpet of beech leaves takes the place of undergrowth. Near this wood stands Oak Hill, an old enclosure in the depth of the wood, which the Lord of the Manor at one time designed for a residence. With an eye to that contingency he planted some of the little valleys with the *Rhododendron ponticum*, and elsewhere a number of Spanish chestnuts. Both kinds have appeared to find a congenial home, and, though I am opposed on principle to the introduction of exotics, these have established their right to the soil. This wood was purchased and thrown into the Forest by Sir T. Fowell Buxton and myself.

I have thus far dealt with the important manors of Epping and Theydon, where there remains very little to do in the way of thinning. Continuing to the south we enter Loughton Manor, in which there is an enormous number of small pollards of oak, beech, and hornbeam, many of which should still be removed. Owing to the fact that the right of pollarding was vested in the whole of the householders in Loughton, no "spear" trees were allowed to develop, and I do not pretend that this is a characteristic which I should altogether desire to maintain. Here an attempt has been made to break the monotony of the wood by clearing considerable patches, which will in time give us young growth, and which incidentally open charming views of such wooded valleys as that known as "Hangboy Slade." I would especially call attention to what has been done in this way, below the green ride which leads from Golding's Hill to Debden Green. In some parts of this manor, as for instance in the neighbourhood of Kate's Cellar, the pollards consist almost entirely of beeches. It should be remembered that these are much more likely to improve, if given room, than hornbeams. Thinning operations are still very urgent in such places.

In some parts of Loughton, Theydon, and Waltham Manors there have been large clearances, of long standing, chiefly caused by fires. In many of these, groves of silvery birches of exquisite grace are filling the gap. Such an one covers the northern slope of the valley below the Loughton Camp. I invite a comparison between it and the opposite side of the valley which is still covered with a monotonous growth of pollards. These, having all been pollarded together, their branches are of the same length, leading to a smooth unbroken outline, as in an artificial plantation, and an absence of shadows.

In the midst of Loughton Manor, but distinct from it in character, lies Monk Wood. Its charm is so well known that I need not further describe it. In 1893 we removed a considerable number of the ugliest of the pollard beeches, a measure which elicited a storm of criticism, but which nevertheless has, in my opinion, greatly improved the wood by giving more scope to the unpollarded beeches. The latter are younger than the pollarded trees, but, not having been subjected to the curtailing process, have far outstripped them by their natural vigour. Observe here the charming groves of young beeches, and hornbeams which have established themselves on the more open parts. The cattle and deer yearly nibble them back, which causes them to hold their brown leaves through the winter. No further removals will be required in Monk Wood, at any rate for a long time to come.

Crossing the main ridge to the north of High Beach, those sections of the Forest, known as St. Thomas's, and Honey Lane Quarters, which slope towards the Lea Valley, are reached. In the former block, which adjoins Woodredden Hill, the hollows are too wet for any trees to grow in them, but they constitute, in themselves, a fresh beauty. Interspersed among the pollarded trees are a considerable number of straight grown beeches, but of much younger growth than those in Epping Thicks and Monk Wood. Having been given plenty of room in the last few years, these trees are making rapid and healthy progress. The upper part of the slope above Honey Lane will, I anticipate, be furnished, in the lifetime of some now living, by a beech grove scarcely inferior to Monk Wood, and its situation is much finer. The exquisite views over Waltham Abbey and the Lea Valley commanded by the whole of this hillside, justify, in my opinion, the large clearances which have been made of the decadent hornbeams which used to cover the whole surface, and had nearly succeeded in destroying the undergrowth. In some portions a few more of them might be removed, but the process is practically complete. Every chance should be given here to the holly which is making progress on the drier banks, so far as the deer, who are very fond of this western slope, and who nibble that prickly plant in winter, will allow. In the lower part of the slope where the soil is wetter, oaks take the place of beeches.

Turning southward again and passing the

"King's Oak" we reach the High Wood at High Beach. This charming old beech wood is of the same character as Monk Wood, except that it is more closely grown, and this has been the case for so long that there is scarcely anything that could be removed from it with advantage.

The Fairmead Thicket covers a wide area of nearly flat ground with a clay soil, and is therefore most suited to the growth of oak and hornbeam, which kinds occupy the ground almost to the exclusion of other species. The wide grass glades bordered by rounded patches of blackthorn are a distinctive and beautiful feature of this district. The oaks, the best of which appear to be of two hundred years' growth, are in some parts too closely ranked for healthy development; but having regard to the difference of opinion that exists on this point, great care will doubtless be exercised in thinning them. Bury Wood, or to quote the name by which it is more commonly known, Hawk Wood, has, for the same reason, excited much controversy. From this wood about three hundred small drawn-up oaks were removed in 1894. The process should, in my opinion, be carried somewhat further, though I am well aware that there are good judges who do not share this opinion. The weak and spindly sticks, to which I refer, in my opinion only help to arrest the development of finer trees, and groups of trees. The chief beauty of an oak lies in its massive lateral branches, but in this wood such laterals are for the most part being killed by overcrowding. A large proportion of the hornbeam pollards associated with them have been taken out, and Hawk Wood will be essentially an oak wood with an undergrowth of thorns. There are, I think, only two beech trees in the whole of it.

Lord's Bushes, a wood of 120 acres, is, I think, in as healthy a condition as any block of the Forest. The fine spear beeches have made great progress in recent years. There are some very rugged and picturesque old hornbeam stems, but large numbers of meaner ones have been cut out. There has been more frequent and more drastic thinning in this grove than in any other part of the Forest. Enough has now been done, I think, at any rate for some time to come. The result of it, and of the fact that there are no commoners' cattle turned out here, is shown in the vigorous growth of holly and of young saplings. But it is the old oak pollards which give it a character all its own, and which cannot be matched in any other part of the Forest. Many of these have been killed by the overshadowing of the lusty young beeches; but others have been taken in time, and saved from the slow but sure process of choking. The excellent foresters who were consulted in 1895, and who are generally referred to as the "experts," seemed to express an opinion in their report adverse to this policy, but I cannot help thinking that there must have been some mis-apprehension, as no one can doubt that the ancient boles to which I refer will continue to give a venerable air to this wood for many centuries to come.

Walthamstow Manor and Gilbert's Slade.—The hornbeams throughout this area are for the most part small and unhealthy; but their place is being taken by fairly active young oak trees, and the holly, which is abundant, should be encouraged. I do not know what is the reason of the absence of beech from these thickets, for much of the ground seems suited to them. Possibly it is due to the idiosyncrasies of some Lord of the Manor. This is a pity, for I think they would stand the showers of blacks from London chimneys, which fall upon the trees in this crowded neighbourhood, better than the oaks will do. The hornbeams being in a moribund condition, these woods will be ultimately composed mainly of oak and holly, as Lord's Bushes is of beech and holly. Much injury has been done to these woods by ill-considered gravel-digging. It is now stopped, I hope never to be revived.

The woods in Wanstead Park are chiefly composed of elms which were planted about one hundred and sixty years ago. There are a few beeches and sycamores among them, and some remarkably well-grown wild cherry trees, as well as some large yews. The old decaying trees on the islands should be preserved as long as possible, as jackdaws and other birds nest in their hollow branches.

The peculiarities of the various woods, which I have endeavoured to indicate, are not confined to the larger growths, but extend to that which covers the surface—grass, heather, brake fern, gorse, broom, or blackthorn—according to the soil and aspect. From the above remarks it will be seen what a charming diversity we have inherited — incomparably more interesting than a wood which, however beautiful, is all of one pattern!

(2) PRESERVATION OF NATURAL FEATURES.—It goes without saying that in a natural forest we should preserve those features which are not of man's doing. As an instance of this may be mentioned the importance of retaining trees which are decaying, trees which are dead, trees which have been overthrown by the forces of nature, as well as those which are in full vigour. I am speaking of course of those of larger growth. I have recently spent a fortnight in exploring one of the largest natural forests in Eastern Europe. Here, to my mind, the chief beauty resides, not in the standing trees, but in the giants that lie prone among their roots. Many of them have lain for several centuries. They are gorgeous with moss and lichen; their great trunks are seed-beds for their descendants, and they tell a story of mighty hurricanes and snowstorms which we should miss, if it had been possible to remove them. Such features are the characteristics of all virgin woodlands. Our Forest is also a document of nature with its tale to tell. Its failures, its ruins, should be preserved, as well as its vigorous youth. It should not be trimmed and garnished.

Of even greater importance is the conservation

and encouragement of all wild life. The Forest is a sanctuary for all birds and beasts, even for those marauders who are generally classed as vermin, unless they can be proved to be so predatory as to endanger other species.

The rule of the Committee on this matter is—"That no shooting be allowed on the Forest unless with the express authority of the Committee, and then only for such animals or birds as are injurious."[1] Besides the protective measures of the Conservators, most of the large proprietors adjoining the Forest have agreed to extend their protection to those birds which are becoming rare, and which especially need it. The birds which they have, up to now, agreed thus to encourage are, the family of owls, all the hawks, except sparrow hawks—and even that little brigand is protected on some estates—the magpie, which is extremely scarce in Essex, the peewit or green plover, the heron, and the kingfisher. In addition to the above measures, powers have been sought by the County Council, and granted by the Home Office, under the Act of 1896, for constituting all the Forest parishes and several adjoining parishes, an area within which bird-catching with nets, etc., is prohibited throughout the year. The only birds which are at present admitted to be injurious on the Forest are the jays, which are there in excessive numbers, and undoubtedly harry other birds' nests; and those rooks who have chosen to breed on the same island as the herons at Wanstead Park. This black-coated colony have so increased of recent years as to endanger the existence of the heronry, with which they are at perpetual war. Fairplay to all things that fly, run, creep, or swim, should be our motto, so that man, who is at times the most predatory animal of them all, may study, and learn to respect them. Nor

[1] There is an exception to this rule which I must mention to prevent misconception. When the Lord of the Manor of Loughton parted with his manorial rights to the Corporation he reserved to himself and to his sons the right of shooting over the Manor of Loughton within the Forest.

should the lower forms of life be forgotten, for these also offer a wide field for profitable research to those who inquire into their secrets; *e.g.* let us leave intact those moist places which teem with life invisible to the passer-by, and in which some budding Hervey or Lister may find his first inspiration.

Experience has taught me to be less hopeful of the flora of the Forest. I have seen the primroses ruthlessly eradicated till there is scarcely a plant left between Epping and London. There is one area where they remain. Let a watchful eye be kept on the Primrose League. Again, the polypody ferns, which used to crown the pollards in Loughton Manor with graceful tracery, have nearly disappeared. I almost despair of a remedy. Some have suggested surrounding portions of woodland with a fence, but let me remind them that such a barrier must be absolutely impregnable. Let those who think the London gamin to be a degenerate animal, watch him, in his leisure moments, climbing, for a pastime, the unclimbable paling which surrounds St. James's Park. Now we want no St. James's Park palings in Epping Forest. We must look to the vigilance of the keepers.

(3) Removal of the Artificial.—The most important and wide-spread interference with the natural aspect of the Forest in the past has been the practice of pollarding. The result, in many parts, is an extremely monotonous condition extending over wide areas. In Loughton Manor, as I have already said, every tree has thus been disfigured. By letting in light and air here and there, we are giving a chance to young trees of recent growth, which will quickly break the smooth outline. I must not be understood to desire a complete removal of all the pollards. Many of them are curious and picturesque, besides which they have a history, and tell of a system of forest management which has had an important place in the centuries past.

To give another instance of the formalities

which have unfortunately been introduced into the Forest, I may mention the hard outlines which form the edges of old enclosures, now thrown open, or of straight rides injudiciously cut through the Forest. It is possible to the experienced forester, who uses his axe as the artist wields his brush, to substitute a perfectly natural and soft outline. I may cite what was done on the edge of the Fairmead Thicket, facing the Forest Hotel, and on the "Clay Road," near Baldwin's Hill, a few years ago—an operation much criticised at the time. I venture to think that that action was justified, and that the criticism was made with imperfect knowledge of the intention of its authors.

(4) Reproduction.—Personally I am opposed to planting on the Forest, except for the purpose of hiding unsightly objects. In the thicket, at any rate, nature is well able to sow her seeds in the most favourable manner, and adequately to protect the saplings. A grove thus produced is infinitely more picturesque and diversified than a planted wood, because the trees are of all stages and ages of growth, and the fittest survive, perhaps after some struggles, of which they bear the signs. An artificial plantation, on the other hand, is, at all stages, even after hundreds of years, distinguishable from a natural wood, chiefly from the reason that the trees are all of the same age and height, and, being grown with nurses, are perfectly straight. I especially deprecate the planting of foreign trees, or abnormal varieties of native ones. Above all, I object to the introduction, into a deciduous forest, of coniferous trees which are not suited either to the soil or the climate. Such exotics at once give an impression of artificiality and of forced culture.

There is a further cogent reason against the multiplication of plantations. Such plantations must be protected with oak posts, which are at present taken from the thinnings of the Forest. The oak woods occupy a very limited area, and have been already considerably drawn upon. Up to the present time such "spears" have been removed for the benefit of the remainder. If more are taken it will be *because they are wanted*—a very dangerous principle to admit in the management of this woodland.

In various parts of the Forest, groves of young trees are coming forward to take the place of more stately trees when these decay. Of course care must be taken that the commoners' cattle are not increased, and that the deer also are kept within reasonable limits, as, if the Forest were "surcharged," no saplings would be allowed to reach maturity. Even now it is only with difficulty that a few "get away." Note how most of them do so only through the protection afforded by thorn bushes. Favourable examples of this may be seen on either side of the road from High Beach to the Wake Arms.

Assistance may be given to the self-sown plantations by a little temporary protection from browsing. Some experiments have been made in this direction by putting a slight fence round certain areas. The weakness of this method is, that though the commoners' cattle are kept out, the deer are too nimble. Possibly a hopeful experiment is to throw down groups of valueless pollard trees with all the branches on, thus protecting the seedlings until they have grown through these branches to a height of four or five feet, but some authorities dissent from this.

For the reasons given above I think it a mistake for the Forest Committee to maintain a nursery. It never can be necessary to plant considerable areas, and the limited number of trees which may be required for the concealment of unsightly objects would be more cheaply bought from nurserymen. Apart from its expense, there are strong reasons against keeping a large stock of young trees. Being there they must be used, and this is a temptation to make more plantations, which are not only no advantage, but a serious eyesore, because they give an artificial "parky" appearance to the Forest. In this opinion, the experienced gentlemen, or at least a majority of them, who were consulted on this and other points in 1894, concurred. Their words are as follows:—"We are opposed to any artificial planting in the Forest," and four out of the five experts signed the following rider:—"As we have expressed a strong opinion against the replanting of the Forest in the ordinary sense, we think there is no need for a nursery, and that it should be discontinued."

THE POLICY OF THINNING.—The leading fact to bear in mind in connection with this question is, that an entirely altered state of things was introduced by the Epping Forest Act, under which the ancient practice of pollarding was abolished. As long as the tops and branches were removed every few years, it mattered little how closely ranked the stems stood. Under no circumstances could trees so treated attain picturesque growth before they were again mutilated. The great difficulty has been to cope adequately with the new conditions involved by the sudden arrest of all pollarding in 1889. From that date the whole of the lopped trees grew on together to their mutual injury, and to the destruction of much that was better worth preserving. Then lateral branches began to die from overcrowding, and in a few years only a few attenuated branches, struggling upwards to the light, would have remained. Moreover, the undergrowth, for want of sunshine and moisture, was disappearing. After twenty years' work a different state of things prevails, and nobody would recommend that the same energetic action to meet the emergency should be continued indefinitely. It is not my purpose to revive the controversies of two or three years ago. I am content to leave it to time to justify what has been done; but for purposes of comparison I will quote what a competent observer wrote in 1864 on the conditions that prevailed in the Forest at that date. Speaking of the privilege of lopping, the writer remarks: "It is continued to this day in certain parishes, and the consequence is that such a thing as a real natural tree is hardly to be found." He then adds: "It is very surprising in such a wood as Loughton, where there are more than a thousand acres of waste, and perhaps a million of trees, to note how not a single one escapes lopping. The visitor from London, walking there in the summer-time, when the lopping is not taking place, and there is nothing to call the villagers into the wood, is struck with the silence and the solitude; but if he begins to indulge any fancies about primeval wastes unspoiled by man, a glance at the trees will correct him. They are not, strictly speaking, trees at all, but strange, fantastic vegetable abortions. Their trunks, seldom more than a foot or eighteen inches in diameter, are gnarled, writhed, and contorted, and at about six feet from the ground, just within reach of the axe, they spread into huge overhanging crowns, from which spring branches which are cut every other year or so, and never long escape the spoiler; then, baffled in their natural instinct to grow into branches, the trees throw up spurs and whips from their roots, and every pollard stump—more or less rotten at the core—is surrounded with a belt of suckers and of sprew. It is no more nature's notion of primeval woodland than are closely-cropped hair and shaven lip and chin her intention for the real expression of the human face." No candid person who remembers the state of things thus described can deny the improvement which is already manifest. Spindly and unhealthy trees have been removed, and promising groups or individuals, relieved from this competition, redouble their vigour, while masses of healthy saplings, holly, and thorn afford shelter to all wild things.

Referring to the suggestions, under the heading of "variety," made above, it will be seen that the area over which active thinning operations are still required is very much narrowed. There is now little that is urgent outside the Manors of Loughton, Waltham, and Sewardstone, and even within these Manors thinning on a greatly reduced scale now meets the case.

In concluding this part of my subject, let me urge that every one who aspires to be a forester should exercise his imagination. He works for posterity. It is essential for his success that he should, with full knowledge of natural processes, have ever present in his mind the probable results of his operations, not only in this generation, but fifty or a hundred years hence.

DRAINAGE.—In this case I am inclined to give Punch's advice to those about to marry. In the Forest, covered drains are sooner or later choked by roots, and open water-courses are quickly trodden in by cattle. No doubt to a certain extent it is necessary to lay dry those tracks which are much frequented; but I think that labour employed in draining swampy hollows is worse than wasted, because, in their natural state, they have a distinct flora, and insect life, of their own. Above all, I deprecate any attempt being made to straighten and formalise the beautiful natural water-courses at the bottom of the hollows. In some cases a perfectly straight-cut ditch has been substituted many years ago for the old water-courses, and the latter may be still traced as a shallow winding hollow. I strongly advocate restoring the stream to its old bed, by excavating it and filling up the straight cut. After a few years the old natural conditions and appearance are restored. This has been done with great success in part of the course of the Ching below Connaught Water, and I hope the process will be carried further.

THE USE OF CARTS.—Great mischief has been done in the past by the use of carts for the purpose of removing gravel and wood thinnings. When a ride has once been cut to pieces in this way, it is years before it recovers, the surface being so ploughed about, that it holds water like a sponge, and only recovers its solidity after a long period. For nearly twenty years I have strongly advocated the plan, which is almost universal on the Continent, of using sledges for this purpose. The runners glide over the surface instead of cutting through it. I am glad to say that in this respect my ideas have begun to prevail, and an order has now gone forth that sledges shall be generally used. There is nothing new in this. It is but reverting to the ancient practice. According to Fisher, there was in 1570 an order that no inhabitants were to draw their customary wood with carts. Later on we find another order that tenants of new tenements should *not* carry wood with a sledge, the meaning of this being, not the prohibition of the sledge, but that these new tenants were only entitled to take as much as they could carry themselves. In 1585 the Rector of Loughton was amerced for carrying away "custome spray," to which he had no right, and forbidden to do so in future under pain of forfeiting "pro quolibet plaustro," *anglicè* sledfull, the sum of twelve pence. In 1718 certain persons were presented for making use of wheel sleds. In 1724 "a sledd with two horses" was, with other particulars, prescribed by the manor court, and a penalty of five shillings to the Lady of the Manor was payable for any violation of this order.

Gravel-Digging.—The Highway Authorities are empowered by the Epping Forest Act to dig gravel after obtaining the consent of the Epping Forest Committee, or if that is refused may apply to the Justices for an order, and this is sometimes granted in the teeth of their opposition. The privilege has been very recklessly exercised in some cases, the conditions laid down by the Forest Authorities being disregarded. A better spirit now prevails. Many of the local authorities have begun to see what a short-sighted policy it is to spoil their own playground for the sake of a doubtful, or at best a slight temporary gain to the ratepayers. The most objectionable feature was the practice of opening numerous small pits close together among the trees or in the middle of some glade or ride. This honeycombing process was due to the fact that the gravel often lies in "pot holes." The pits being narrow and deep were dangerous, and an attempt was generally made to fill them up partially by sloping the sides to resemble the inside of a pie dish, and by filling them with the top spit removed from the roots of the trees for many yards around. The ultimate result of this process was a noxious black swamp with unpleasant exhalations. Where gravel-digging is still allowed, the pits should be continuous, even if a few inferior pollard trees are thereby sacrificed, and above all no attempt should be made to formalise their steep edges, the deeper and more rugged they are left the better will they ultimately harmonise with their surroundings. I do not object to an extensive gravel-pit in a suitable position left with its original rough edges and clothing itself anew with gorse and broom in the shade of which lies a clear pool of water. Such an opening in the Forest is a beautiful feature, and very different from the destructive delving against which I protest.

GAMES.—There is a remarkable increase in the number of football and cricket clubs who apply for the use of grounds on the Forest. The taste for athletics grows apace, and I trust the Committee of Management will always endeavour to satisfy this aspiration as far as possible. Large amounts have been spent at Wanstead Flats by outside associations, like the London Playing Fields Committee, on the preparation of grounds. I hope that the Conservators will not be content with an attitude of benevolent neutrality towards such associations, but will actively promote them. Even now no more beautiful sight can be seen on the Forest than when thirty or forty cricket matches, comprising some five hundred cricketers, are being played at once on a Saturday afternoon. Chingford Plain, though somewhat farther from London, will obviously be utilised in the same way before long. The golf players are already in possession, but it ought not to be difficult to define the respective spheres of influence, the more hilly parts being reserved for golf, the flatter portions for football and cricket.

Forest Fires.—This scourge is responsible for much mischief and many unsightly clearances which were unjustly attributed to the recklessness of the Conservators. Some are the result of carelessness; some of thoughtless larking. A few, I fear, are caused by deliberate incendiarism. In the latter category must, I fear, be placed the disastrous fires which occurred in 1894 in St. Thomas's Quarters and its neighbourhood. The most inflammable material is dead heather and brake fern, and the most dangerous period is during dry weather in March and April. Owing to the fact that the vessels are then full of rising sap, the injury, which in the case of young trees means death, is then most marked. Something may be done by raking away inflammable growth from the base of the stems of young trees, but the best safeguard is, whenever a dry spell occurs in the spring, to multiply the watchers in those parts which are in the most critical condition. It is worth noticing that birch trees, of twenty years growth or more, enjoy practical immunity from the effects of fire owing to the roughness of bark which they develop at the base. Oaks have the same valuable quality, but they only attain it at a much later stage of growth, and in a young state are peculiarly susceptible. The older beeches are protected by the fact that surface growth does not thrive under their branches.

Boundaries of the Forest.—I regret to say that small encroachments on the boundaries of the Forest still sometimes occur. It seems almost unnecessary to point out that Trustees of public lands are under an obligation to be at least as strict in such matters as private owners. Above all, should they themselves be above suspicion of showing favour to individuals, or of enclosing a yard of forest land for any purpose except as authorised by the Epping Forest Act. It has been contended that the Conservators have the right to enclose portions of the Forest adjoining the superintendent's residence, for its protection, but I can find no authority for this in the Act of Parliament, which is extremely explicit as to the purposes for which forest land may be enclosed, and expressly forbids enclosure for any other reason. The official map on the scale of 25 inches to the mile is the paramount authority as to the actual boundaries. An idea prevails that where a hedge and ditch bound the Forest, both belong to the adjoining proprietor, but this rule is subject to many limitations. I advocate a series of permanent boundary posts or marks, wherever there is room for question. This was the practice in olden times.[1]

SEWAGE.—It seems almost an offence even to name the word in connection with the Forest. Yet there are people who, because it is technically described as "waste," think that they are entitled to dump down their filth on it. After twenty years of protest against this outrage, there are still cases in which abominations are allowed to flow on to it as if it were no man's land. I admit the difficulties in which isolated houses may be placed. Yet the resources of civilisation have not been exhausted. Whenever there is a bad outbreak of cholera we shall perhaps succeed in effecting a complete reform.

OLD CULTIVATION MARKS.—Large blocks of the Forest were enclosed about 1860, the trees removed, and the land ploughed up. Although they have been long ago restored to the use of the public, the "ringes," as the cultivation ridges are called, still remain. They are very uncomfortable to ride or walk upon, a hindrance to games, and ugly in themselves. Some experiments have been made with a heavy steam roller. They were expensive, but fairly successful, when the weight was persistently applied, and the land in the proper condition. I hope the Committee will some day try it on a larger scale.

[1] Professor Meldola has recently investigated the subject, and has succeeded in unearthing several of the old boundary stones of Waltham Forest.

COMMONERS' RIGHTS.—One of the most thorny questions which the Conservators have to deal with is the ancient right of pasturage. As the preservation of the Forest turned upon commoners' rights it would ill become the Conservators to make any encroachment upon them. On the other hand, it is obviously their duty to see that the Forest is not "surcharged" by commoners' cattle. In the interest of the commoners themselves, and also to preserve the thicket from undue trampling and browsing, the number of cattle pastured should not be increased to the starvation point. The old theory and practice of the Courts of Attachment, that every rightful commoner should have some enclosed land in his own occupation, whereon to maintain his cattle during the fence month, is a useful principle to adhere to, even at the present day, because, although the fence month is not likely to be enforced, there is very little available feed in the winter, and unless the cattle which can really pick up a subsistence in the spring and summer have other support in the scarce time, they would inevitably, in the struggle for food, seriously injure the young trees, while their emaciated forms would be painful to contemplate.

It is roughly computed that in an average summer the outside number of cattle, for which there is sufficient feed, does not exceed 800 or 1000. It is on this account that it has been determined that the minimum area of land which entitles to a commoner's right should be half an acre. At the same time, as it is found that a large proportion of the commoners do not exercise their right, the Conservators annually allow to a number of poor cottagers—about 170 at the present time—the privilege of having two cows or one horse marked by the Reeve, and depastured as though they were commoners. It is desirable on the one hand that these cottagers, who have previously exercised this privilege, should not be deprived of it, and on the other it is in their interest, as much as of any one else, that there should be no laxity, or undue extension of this boon. Those who desire to inquire more particularly into the technicalities of this subject, I refer to the Epping Forest Act, sec. 33, Fisher's *Forest of Essex*, chap. vi., and the well-known judgment of the Master of the Rolls.

APPENDIX

MONK WOOD, IN LOUGHTON

A FRAGMENT OF FOREST HISTORY[1]

By W. C. Waller, M.A.

The following very interesting note on the history of Monk Wood by that excellent authority on old documents, Mr. W. C. Waller, which I am permitted to extract from the *Essex Naturalist*, throws no little light on contemporary customs, and is worthy of study:—

"Somewhere about the beginning of the thirteenth century one part of Loughton was called 'Luketon Snarringe,'[2] as being, or having been, the fee of one Geoffrey de Snarring. He, we may say in passing, was probably an under-tenant of the great Norman barony of de Valoines in Essex, as in Norfolk; from a place in the latter county, now called Snoring, he seems to have derived his name. It appears, however, that he had granted at any rate some portion of his estate to three owners, who held a certain wood in Luketon Snarrynge in common, though their shares were not equal. But at this point it will be well to let two of them, Geoffrey Renitot and Roger Fitz-Ailmar, speak for themselves through the medium of an interpreter, their own words being recorded in monkish Latin:—

"To all the faithful in Christ, Geoffrey de Renitot and Roger Fitz-Ailmer send greeting in the Lord. Be it known

---

[1] *Authorities:* Harl. MS. 4809. D. of Lanc.: Surveys and Depositions; 24 Eliz. D. of Lanc.: Misc. Rec.; xxv. F. 17a.

[2] It may have been a small manor. That a hill on the southern border of Monk Wood is still called "Court Hill" is a significant fact.

unto all men that Ralph de Assartis and we ourselves having measured of our commonwood in the parish of Loughton, fifty-six acres and a-half, and Ralph having demised his share to the Abbot and monks of Stratford,[1] we, for our own salvation and that of our [kinsfolk], granted, gave, and by this charter have confirmed, to God and the Church, and to the Canons Regular of the Holy Cross of Waltham, all our part of the aforesaid wood with the land and all the right we had therein, or could have, as in the felling and carrying-away of trees, and in pannage at pannage-time, with everything else by any right pertaining to us, to be had and held by the said church and canons in free, pure, and perpetual alms, free and quit from all secular service or demand from us and our heirs for ever. And it is to be noted that of the aforesaid wood and land, fifty-six acres and a-half by measure, our share was a fourth part in all the advantages mentioned, and others which might casually accrue, to be received in common between the said Ralph de Assartis and ourselves. And we and our heirs will for ever guarantee against all men this part of the wood and land, with the appurtenances thereof, as is aforesaid, to the said church and canons, as our free, pure, and perpetual alms. And for this grant, donation, and the confirmation of this our charter, the aforesaid canons have received us for ever in the prayers and other benefits of their house. These being witnesses, etc.

"Not content with this joint declaration of their gift, Geoffrey and Roger proceeded to execute, each of them, separate deeds, couched in much the same terms. The names of four witnesses are appended to the charter of the former, viz. Nicholas de Barton (Barrington), William de Bosco, Richard Alcher, and Gregory de Thayden. The last is probably identical with the man of that name who was a verderer in 1250 A.D.

"Nor was this the conclusion of the whole matter. Following on the three charters just recited we have two others, from which we learn that both Roger and Geoffrey still had seventeen acres of wood and waste left to them, of which Roger's share was three acres and three-parts of a rood (*rode*), and this they also made over to the canons. The grant made by Geoffrey was subsequently confirmed

[1] Unfortunately no cartulary of the Abbey of Stratford Langthorne, if it exist, is accessible. That said to be preserved in Trin. Coll., Dublin, is a fragment of three pages only.

by Edward, his son. A pleasing unanimity has marked the proceedings up to this point; but the new joint-possessors do not seem to have succeeded in maintaining it. For, on the Thursday next after the Feast of Saint Barnabas the Apostle, in the twenty-fourth year of the reign of King Henry, the son of King John (14th June 1240 A.D.), Henry, Abbot of Waltham, and Hugo, Abbot of Stratford, found it desirable to meet in the mother church of Chelmsford and there to compose certain differences which had arisen over their common wood in Loughton Snarrynge. The result of their meeting is recorded in a charter, by which it is solemnly provided that, when either Abbot wishes to fell any timber, the bailiff of the one shall send for the bailiff of the other, and the two shall, by common consent, fix upon four trees of equal value, of which the Abbot of Stratford shall have first choice as to two, and the Abbot of Waltham take whichever he prefers of the two remaining. Into the other provisions we need not enter here.

"So far so good, says a logical reader: we have a wood and we have monks; but there is nothing to show that the wood was called 'Monk Wood'; nor even if it were so called, that it was the particular wood which now goes by that name. To meet these objections, which are reasonable enough, we must carry the reader from the thirteenth to the second half of the sixteenth century, when Elizabeth, by the Grace of God, was Queen, and, withal, lady of the manor of Loughton, *alias* Lucton.

"From a Commission to survey, dated 20th May 1582, we learn that 'greate spoyle and waste' was alleged to have been committed in the felling of a parcel called 'Moncke Wood,' parcel of the Manor of Loughton, lately sold to Robert Wroth, Esq., by Thobie Houghton, surveyor of woods to the Duchy of Lancaster. The three commissioners named were directed to repair to Mouncke Wood, then and there calling before them Robert Wrothe and others. Their report states that they did so on 1st June, and 'the same daie at Eppinge, did by the othes of . . . . twelve substanciall and honest men, neare inhabytinge to the said mannor of Loughton, make inquirie of the same, who, uppon viewe as well made by them of the woodes as by seekinge further to understand of the same . . . have made presentment.'

"The presentment, or verdict, consists of detailed answers to five articles of inquiry; and as they are brief and to the point, we give them as they stand:—

"(i.) We say that there is a wood uppon the waste soyle of the said mannor called Muncke Wood, containing as it is measured fifty-three acres, sixty-five poles, at twenty-one foote to the pole; whereof there is waste ground in the same that beareth no wood by estimacion fifteen acres; which said wood hath been sold to Mr. Wroth, who felled the same. The nature and kind of the woodd so felled was most oke, beach, hornebeame, and birch. The oaken wood was lopte and some shredde, and the other usual wood was most lopte, saving there was felled by the ground of the said usuall wood to the nombre of 500 younge trees. And as we are certifyed by our evidence, it hath byn felled in lyke order before at former sales. The said wood at the time of the fellinge thereof was fifteen years' growthe.

"(ii.) We say that there was late felled within the said wood eight timbre trees for making of a pownde at Loughton Hall; which is informed to be done by warrant from Mr. Chancellor of the Duchy. Of crabtrees and hawthornes, to the number of 618 trees, and two hollies being vert. And as we are informed moste parte of them were dead in the toppe and felled by carters and beaten down on the fall of the wood.

"(iii.) We say that Robert Wrothe, Esq., paid for the same wood to the Queen £20. The charges of felling, etc., stood him in £35; and he afterwards sold the said bargaine of wood to Philip Grenely for £120, giving him one year and a quarter's daie for paiment of £90. And Philip Grenely saith uppon oath that he got not £20 by the said bargaine.

"(iv.) We say that the verte felled in Muncke Wood aforesaid was felled by the foresaid Philip Grenely between the Feast of St. Bartholomew and the Feast of St. Michael last past, after Mr. Wrothe had sold the bargain of woode to him; but whether the doing thereof is to be accounted waste or not, we knowe not.

"(v.) We say that the said Munckewoodd hath byn three times sold within the mynde of man: that is, one tyme by the Abbot of Stratford, and twice in the Queene's Majestie's tyme that now is.

"In witness of this our Verdict we have hereunto sett our handes and seales the xii. June, 24 anno R.

"Two points are to be specially noted in the foregoing report. First, the area (fifty-three acres, sixty-five poles) of the wood; and, secondly, the fact that it was sold, presumably before the dissolution of the monastery, by the Abbot of Stratford. A search through one or two *Ministers' Accounts* for proof that Monk Wood was included among the possessions of the 'late dissolved monastery of Stratford Langthorne,' proved vain; nor does it seem to be numbered, with the manor of Loughton, among those of Waltham Holy Cross. But woods were apparently entered on the rolls only when the proceeds of their sale came into the accounts. Of a great felling which took place in or about A.D. 1488, we have evidence in a Forest Roll (4 Hen. VII.), according to which a certain Christopher Stubbes, of Loughton, was presented for having cut down 100 loads of timber and wood in Monk Wood, called 'le King's wast soile,' and for the bark of the same received viij. s. The explanation of this would seem to be that the Abbot had sold the wood to Stubbes, without first obtaining a licence to fell.

"The evidence, however, seems sufficient to warrant the identification of the Abbot's Wood in Luketon Snarringe with that now known as Monk Wood, in Loughton."

THE END

Printed by R. & R. CLARK, LIMITED, *Edinburgh*.

Spotted something that isn't quite right?

Let us know and we'll fix it for future reprints. And as a thank you for your diligence, we'll send you a free Sothis book of your choice.

publisher@sothis.press

This page intentionally left blank

Made in the USA
Monee, IL
08 November 2026

67979636R00121